S. Hrg. 114–474

THE STATUS OF ADVANCED NUCLEAR TECHNOLOGIES

HEARING

BEFORE THE

COMMITTEE ON ENERGY AND NATURAL RESOURCES UNITED STATES SENATE

ONE HUNDRED FOURTEENTH CONGRESS

SECOND SESSION

MAY 17, 2016

Printed for the use of the
Committee on Energy and Natural Resources

Available via the World Wide Web: http://fdsys.gov

U.S. GOVERNMENT PUBLISHING OFFICE

22–127   WASHINGTON : 2017

For sale by the Superintendent of Documents, U.S. Government Publishing Office
Internet: bookstore.gpo.gov   Phone: toll free (866) 512–1800; DC area (202) 512–1800
Fax: (202) 512–2104   Mail: Stop IDCC, Washington, DC 20402–0001

COMMITTEE ON ENERGY AND NATURAL RESOURCES

LISA MURKOWSKI, Alaska, *Chairman*

JOHN BARRASSO, Wyoming
JAMES E. RISCH, Idaho
MIKE LEE, Utah
JEFF FLAKE, Arizona
STEVE DAINES, Montana
BILL CASSIDY, Louisiana
CORY GARDNER, Colorado
ROB PORTMAN, Ohio
JOHN HOEVEN, North Dakota
LAMAR ALEXANDER, Tennessee
SHELLEY MOORE CAPITO, West Virginia

MARIA CANTWELL, Washington
RON WYDEN, Oregon
BERNARD SANDERS, Vermont
DEBBIE STABENOW, Michigan
AL FRANKEN, Minnesota
JOE MANCHIN III, West Virginia
MARTIN HEINRICH, New Mexico
MAZIE K. HIRONO, Hawaii
ANGUS S. KING, JR., Maine
ELIZABETH WARREN, Massachusetts

COLIN HAYES, *Staff Director*
PATRICK J. McCORMICK III, *Chief Counsel*
BENJAMIN REINKE, PH.D., *Congressional Fellow*
ANGELA BECKER-DIPPMANN, *Democratic Staff Director*
SAM E. FOWLER, *Democratic Chief Counsel*
RORY STANLEY, *Democratic Legislative Aide*

(II)

CONTENTS

OPENING STATEMENTS

WITNESSES

ALPHABETICAL LISTING AND APPENDIX MATERIAL SUBMITTED

THE STATUS OF ADVANCED NUCLEAR TECHNOLOGIES

Tuesday, May 17, 2016

U.S. SENATE,
COMMITTEE ON ENERGY AND NATURAL RESOURCES,
Washington, DC.

The Committee met, pursuant to notice, at 10:05 a.m. in Room SD–366, Dirksen Senate Office Building, Hon. Lisa Murkowski, Chairman of the Committee, presiding.

OPENING STATEMENT OF HON. LISA MURKOWSKI, U.S. SENATOR FROM ALASKA

The CHAIRMAN. Good morning. The Committee will come to order.

Senator Cantwell will be joining us, but I understand she is stuck behind a motorcade.

We come to order this morning to begin the hearing on the status of innovative advanced nuclear technologies. We are holding this hearing because nuclear energy must be a national priority. It provides about 20 percent of our nation's electricity and 63 percent of our emissions-free electricity. It is safe, and it is extremely reliable.

When cold winters hit the Northeast and the flow of natural gas is restricted, nuclear plants can continue to provide electricity to residents, literally saving lives. When the wind is not blowing and the sun is not shining, nuclear is still providing essential base load capacity. For any number of good reasons, nuclear has to remain part of the energy mix.

In addition to supporting the current nuclear fleet, I have long supported the research, development and deployment of next generation nuclear technologies, including small modular reactors, micro-reactors, Generation IV reactors and future fusion reactors. That support appears to be growing here in Congress which is a good thing for our country.

We are entering a new era for nuclear power. The opportunity for innovation in nuclear technologies has not been this great since the 1960's. Despite the many difficult challenges associated with full deployment, technical—financial, bureaucratic and license-related—there is unprecedented interest from both the public and private sectors. We can help by removing barriers and optimizing our public-private partnerships.

Despite the clear benefits of nuclear energy, the industry is at a crossroads. The current operating fleet faces a number of challenges due to political decisions, or state market designs and regulations that skew the value of nuclear to the grid. Moreover, the

(1)

President's greenhouse gas regulations do not value the contribution of nuclear-generated electricity on a level footing with the other sources of emissions free electricity.

In order to facilitate the emergence of advanced nuclear technologies, we adopted the Nuclear Energy Innovation Capabilities Act, which was sponsored by Senator Crapo, as an amendment to our broad, bipartisan energy bill. Senator Crapo is with us this morning. Thank you, Senator, for your contributions on this important issue. That amendment was adopted by a vote of 87 to 4, highlighting the value placed on nuclear innovation on both sides of the aisle. The House has passed a nearly identical bill which we hope, when signed into law, will be a valuable step in getting advanced nuclear technologies to the market.

As these technologies are developed they will face further challenges. To give one example, they will have to navigate a complicated and expensive NRC licensing process as they come closer to deployment.

I am pleased to have joined, as a co-sponsor, Senator Inhofe's bipartisan bill, the Nuclear Energy Innovation Modernization Act. In many ways that bill compliments the provisions within our energy bill. It helps reform the NRC in smart ways without compromising safety, and I am hopeful that it will be reported quickly out of the Committee on Environment and Public Works.

Beyond new legislation we must also continue our fiscally responsible support for nuclear research and development. In that vein, I was pleased that we were also able to increase the authorizations for both the Office of Science and ARPA-E within our energy bill.

I believe that removing bureaucratic barriers to public-private partnerships, reforming the licensing structure, and continuing responsible funding for nuclear science RD&D will help drive these innovative technologies to revolutionize the industry and provide robust economic growth.

Our nation deployed the first commercial nuclear power plants, and our regulatory structure is still considered the gold standard. Our universities and national labs are world leaders in education and research. I see advanced reactors as the next chapter of America's leadership in this field.

This is critical because we must remain the go to country for nuclear know how, especially as many foreign nations increase investment and try to challenge our dominance in this industry. Our nation must continue to be the major player on the world stage for nuclear energy, and we must be able to deploy our innovative advanced reactors here at home.

I am pleased to welcome our very esteemed panel of witnesses. We have representatives from a great cross section of advanced reactor technologies at different stages of commercialization, a national laboratory that has been a nuclear leader for decades, and a utility that has consistently supported current and advanced nuclear technologies. I welcome all of you here today.

With that I will turn to Ranking Member Cantwell, welcome.

STATEMENT OF HON. MARIA CANTWELL, U.S. SENATOR FROM WASHINGTON

Senator CANTWELL. Thank you, Madam Chair, and thank you for calling this important hearing on advanced nuclear technology.

I would also like to thank the panel for being here today and for their work in this important field. The group of witnesses will provide us with a comprehensive viewpoint needed to explore the current state of advanced nuclear technology in this country.

Nuclear energy has provided nearly 20 percent of electrical generation in the United States over the past two decades and currently produces 60 percent of American's carbon free-electricity, but the 99 reactors licensed to operate today in the U.S. will not last forever. If nuclear power is to remain part of our energy future, we need to develop and demonstrate the next generation of nuclear power.

For me, we also need to deal with the challenges of nuclear waste. I should say, we all need to deal with this. It is something that presses every day for us in the State of Washington.

The lack of a comprehensive set of solutions has hampered both commercial nuclear development as well as our defense waste cleanup efforts in this country. Secretary Moniz has worked hard to break the log jam, and I think this Committee will ultimately play a key role in crafting a path forward on this very technically challenging issue.

Meantime, we should also acknowledge that while nuclear power has a record of operating safely and cost effectively, there is also potential for catastrophe like we have seen at Fukushima, Three Mile Island and Chernobyl.

If nuclear power is to have a future, the problems that we have consistently been plagued by in the past must be met with innovation and effectiveness. New designs must be safer, cheaper and efficient, and proliferation resistant. In addition, we must have licensing and regulatory systems that ensure nuclear power is not only safe but accepted by the public so transparency and open communication by the industry and government is also important.

The Department of Energy and private industry have been working to address these problems. There are several designs being considered here in the U.S. and globally that have promising features to address some of these long standing issues. I look forward to hearing from the witnesses today on those specific technologies.

Advanced nuclear may, someday, make a real contribution to advanced manufacturing in Washington State and the Northwest Region. I am pleased to have here TerraPower, NuScale and the Idaho National Laboratory. The Pacific Northwest National Laboratory is also making important contributions in advanced nuclear development.

The Northwest has proven to be an exciting place for the development of advanced nuclear technologies, and NuScale Energy, Energy Northwest, Utah Associated Municipal Power Systems are all partnering to construct and operate the country's first small modular reactor.

In addition, TerraPower and the Chinese National Nuclear Corporation signed a memorandum of understanding to develop TerraPower's traveling wave reactor. So it is clear that making a dent in the global carbon emissions will require cooperation between U.S. and China which may prove to be one of the biggest nuclear-energy markets as well as a testing ground for advanced nuclear.

TerraPower's engagement with China is an important example being set for advanced reactor technologies. In order to be a part of the new wave of nuclear energy, the U.S. must be a strong exporter of advanced proliferation resistant nuclear materials and technology.

So the advancement of nuclear technology is an important pathway for the global community to move away from carbon emitting technologies. It is vital that the U.S. continue to lead in this area of clean energy, and nuclear solutions may prove to be a key component of our overall efforts.

Again I thank the Chair for holding this important hearing, and I look forward to hearing from the witnesses.

The CHAIRMAN. Thank you, Senator Cantwell, and thank you to our witnesses. We attempted to schedule this earlier in the year and unfortunately you got bumped. Thank you for your flexibility and your willingness to come back before the Committee.

At this time, I will introduce a few of the witnesses, and we have a couple members here who would like to do more detailed introductions of some of our witnesses this morning.

Our panel this morning will be lead off by Dr. Jacob DeWitte, who is the Co-Founder and the CEO of Oklo, welcome this morning.

Dr. John Gilleland, who is the Chief Technical Officer at TerraPower, which Senator Cantwell has just mentioned.

Mr. Hopkins will be introduced by Senator Daines this morning, as we understand that he is a fellow Montanan, but I also know that he is a pretty strong fisherman that comes up to Alaska occasionally. So, welcome to the Committee.

Senator Daines?

STATEMENT OF HON. STEVE DAINES, U.S. SENATOR FROM MONTANA

Senator DAINES. Well it is my honor to have John Hopkins here from Superior, Montana. He is the Chairman, CEO of NuScale, and it is great to have him with us here today.

I am very much looking forward to his testimony and very excited about the innovation coming out of this group of panelists, specifically from John Hopkins' group, on these modular nuclear reactors as part of the solution going forward here with our all-of-the-above energy portfolio.

Welcome, John.

Mr. HOPKINS. Thank you, sir.

The CHAIRMAN. Thank you, Mr. Hopkins, for being here.

After his testimony we will welcome Steve Kuczynski to the Committee. He is the President, CEO and Chairman of Southern Nuclear Operating Company, welcome. It is good to have you here.

And Dr. Peters will be introduced by Senator Risch.

STATEMENT OF HON. JAMES E. RISCH, U.S. SENATOR FROM IDAHO

Senator RISCH. Thank you very much, Madam Chairman.

Dr. Peters, welcome.

I would like to introduce Dr. Peters, who is relatively new at the Idaho National Laboratory.

Dr. Peters comes to us from Argonne, where he was an Associate Lab Director, and prior to that he had been employed as a scientist at Los Alamos. He has a deep background in nuclear energy.

It is fitting that he comes to Idaho which is the flagship and lead nuclear energy laboratory in America, which we are very proud of. And that is for good reason. Most people do not realize that the first light bulb lit by civilian energy from nuclear power happened in Idaho at the Idaho National Laboratory. We are very proud of that, and we maintain that position.

So, Dr. Peters, we are glad to have you. I am sure we will see you frequently here at this Committee.

With the Chairman's permission I would like to introduce my colleague and friend, Senator Crapo.

Senator Crapo, although not on this Committee, grew up in the shadow of Idaho National Laboratory in Eastern Idaho. He remains engaged and interested. He and I partner on virtually everything that we do over there. In fact, right now we are partnering on a couple of pieces of legislation which we just dropped. With the Chair's permission, I would like Senator Crapo, maybe, to just give us a couple of sentences on those two pieces of legislation which we are introducing.

Mike?

The CHAIRMAN. We are happy to invite you before the Committee, Senator Crapo, and I am happy to have had an opportunity to join you all out at the Idaho National Lab.

Senator RISCH. That is true. That is right.

The CHAIRMAN. Very, very informative, and a very necessary, a very necessary, tour for so many of us.

Senator Crapo, if you would like to say just a couple of words?

Senator CRAPO. I know this is not the usual procedure, so I will be very quick——

The CHAIRMAN. I do not think you are being picked up by the record.

Senator CANTWELL. We are happy if you sit on our side of the dais.

Senator RISCH. Yes. [Laughter.]

The CHAIRMAN. There you go.

Senator RISCH. I can assure you he is not very comfortable over there. [Laughter.]

Senator CASSIDY. Can we get a photo of that? [Laughter.]

Senator CRAPO. No, not with the sign. [Laughter.]

STATEMENT OF HON. MIKE CRAPO, U.S. SENATOR FROM IDAHO

Senator CRAPO. Well thank you very much. I know this is unusual. So thank you, Senator Risch and Madam Chairman, for allowing me to just say a few sentences.

Actually, Madam Chairman, you identified the legislation that Senator Risch just referenced in your introductory comments. So I won't elaborate further on that.

We've got NEIMA and NEICA, the two major bills that will reform both the process and create the new emphasis for going into our new advanced nuclear reactors and helping to make them much more aggressively facilitated.

I am excited about all of this, and I look forward to working with all of you. Thank you for letting me, kind of, join the Committee for a moment.

The CHAIRMAN. We appreciate it. Thank you for your leadership on this, Senator Crapo.

With that, let's begin with our panel of witnesses.

Dr. DeWitte, if you would like to lead off, please?

STATEMENT OF DR. JACOB DEWITTE, CEO AND CO-FOUNDER, OKLO INC.

Dr. DEWITTE. Thank you.

Chairman Murkowski, Ranking Member Cantwell and distinguished members of this Committee, I want to thank you for holding this hearing and for giving me the opportunity to testify. I'm honored to be here today, and I'm excited that you are holding this hearing because I've been passionate about nuclear technology since my childhood.

I was born and raised in Albuquerque, New Mexico, where my Saturdays as a young boy were often filled with my father taking me to get donuts followed by a visit to the National Nuclear Science Museum. And during those trips I recall being captivated by the science and technology and physics of nuclear power, and I knew from a young age that I wanted to work on nuclear reactors. So I am Jacob DeWitte, the CEO and Co-Founder of Oklo. Oklo is a Silicon Valley-based company developing a very small, advanced reactor that produces two megawatts of power. That's very small relative to without the nominal size of a thousand megawatts. We like to call it sometimes a micro-reactor or a nuclear battery and it is designed to bring distributed, clean, affordable and reliable power in small packages to the market.

These reactors fit into containerized systems that can power a wide variety of markets both domestically and internationally which do not have access to affordable and reliable power and in some cases, do not have access to power at all. Our reactor operates purely on natural forces with very few moving parts in the entire system, and it is designed to operate for 12 years before refueling. It will produce reliable, affordable, safe, emission-free power wherever it is needed and the reactor is sized appropriately to open up new opportunities for clean and safe nuclear power in remote and rural communities as well as industrial and military sites in areas that are too small for larger reactors.

The Oklo reactor has the potential to reduce these customers' energy bills by up to 90 percent. Furthermore, our reactor is up to 300 times more fuel efficient than current reactors and can actually consume the used fuel from today's reactors as well as depleted uranium stockpiles around the nation. In fact, our reactors and others like them, could power the world for 500 years with the global inventory of used fuel and depleted uranium, all while reducing the radioactive lifetime of those materials.

Our reactors can also assist with plutonium disposition by consuming excess cold war materials and turning them into clean, peaceful energy.

We started Oklo because we believe advanced reactors will be a significant part of the energy mix of the future, and we wanted to make that future a reality as quickly as we could.

So advanced reactors can provide clean, affordable, reliable and extremely safe, carbon free power that can be deployed on a global scale. They offer the promise to realize the energy future envisioned by the intellectual giants upon whose shoulders we all stand. Fermi, Weinberg, Wigner, Seaborg, and others all saw the potential that next generation reactors have.

Some of the key attributes of advanced reactors include a competitive economics due to reduced capital cost and shortened construction times, multiple energy output streams ranging from electricity to process heat, improved fuel efficiency and the ability to consume used nuclear fuel, flexible operations such as load following and grid stabilization to couple with renewables, and passive inherently safe designs producing walk away, safe technologies as well as flexible siting independent of access to cooling water. Additionally, advanced reactors enable a broad diversity of reactor sizes. Micro-reactors like ours can bring affordable and reliable nuclear power to areas that cannot support larger plants. Alaska and Hawaii are good examples.

But there are a number of places in the continental U.S. as well as other U.S. territories that are excellent candidates for these reactors. The size and characteristics of our reactors also enable us to reach markets that are underserved by existing energy technologies.

Looking farther afield, advanced reactor technologies can fuel mankind's ambitions of navigating the stars. We need energy to explore the heavens and nuclear energy will power future trips to our neighboring planets and beyond. This is not science fiction. This is work that is actually happening today.

Dozens of startups and large companies are now working to commercialize advanced reactor technologies in the United States. Nuclear innovation is alive and well and advances in computational simulation and modeling, along with an injection of talented, young, creative, hungry engineers into the nuclear industry have fueled much of this growth. Federal efforts to attract students into nuclear engineering programs over the last decade are paying dividends and there is more to come.

This activity has also attracted over $1 billion in private investment. And these investments are supporting advanced reactor companies because of their massive market potential as well as the environmental benefits of next generation reactors. And while the capital invested so far is significant, there is still much more that can and will be invested in advanced reactor projects, especially if some of the remaining hurdles are cleared.

Advanced reactor developers face a variety of hurdles and challenges to deploying their technologies. One such challenge is the regulatory process which is significant and necessary challenge, but it's a challenge that advanced reactor developers must navigate.

Unfortunately, the regulatory process, as it exists today, is not a good fit for these new technologies and the venture finance models that fund them. On the other hand, I must emphasize that the widely held belief that advanced reactors cannot be licensed today is also mistaken. We have found clear licensing pathways for our technology, but at the same time there is room for significant improvement and modernization.

We laud the recent work done by the Department of Energy and by this Committee and by the Senate as a whole both supporting the recent legislation that passed into the Energy bill as well as the pending legislation for regulatory reform. These are crucial steps to help us seize the tremendous opportunities in front of us, to advance nuclear power and also the massive opportunities that we have to be the leader at the global stage.

Thank you.

[The prepared statement of Dr. DeWitte follows:]

Testimony of
Dr. Jacob DeWitte
CEO and Co-Founder, Oklo Inc.

Before the Committee on Energy and Natural Resources
United States Senate

Status of Advanced Nuclear Reactor Technologies

May 17, 2016

Written Testimony

Chairman Murkowski, Ranking Member Cantwell, and distinguished members of this Committee, I want to thank you for holding this hearing and for giving me the opportunity to testify. I am honored to be here today.

I am Jacob DeWitte, co-founder and CEO of Oklo Inc. We are a Silicon Valley based company developing and building a very small advanced reactor that produces 2 MW of electric power and is designed to bring distributed, clean, affordable, and reliable nuclear power to the market. These reactors fit into a containerized system that can bring power to a wide variety of markets both domestically and internationally, which do not have access to affordable and reliable power, and in some cases, do not have access to power at all. Our reactor operates purely on natural forces, with very few moving parts in the entire system, and is designed to operate for 12 years before refueling. It will produce reliable, affordable, safe, emission-free power wherever it is needed. The reactor is sized appropriately to open up new opportunities for clean and safe nuclear power in remote, rural, and native communities, as well as industrial and military sites in areas that are too small to for larger reactors. The Oklo reactor has the potential to reduce these customer's energy bills by up to 90 percent. Furthermore, our reactor is up to 300 times more fuel efficient than current reactors, and can actually consume the used fuel from today's reactors, as well as the depleted uranium stockpiles around the nation. In fact, our reactors, and others like them, could power the world for 500 years with the global inventory of used fuel and depleted uranium, all while reducing the radioactive lifetime of those materials. Our reactors can also assist with plutonium disposition by consuming excess Cold-War-era materials and turning them into clean energy for peaceful purposes.

I am excited about today's hearing because I have been passionate about nuclear energy since my childhood. I was raised in Albuquerque, New Mexico, where my father often took me to the National Atomic Museum (now known as the National Nuclear Science Museum). I was amazed by nuclear science and technology, and our ability to get so much out of something so small. I believe advanced reactor technologies can fulfill the promise of the field, and I am proud to be able to contribute to the effort to bring these extraordinary technologies to market.

Advanced Reactors

Advanced reactors can provide affordable, carbon-free electricity and industrial process heat at a global scale, while also providing a new option to keep America's nuclear energy fleet strong as existing reactors reach the end of their licensed lifetimes. Advanced can realize the energy future envisioned by the intellectual giants upon whose shoulders we all stand: Fermi, Weinberg, Wigner, Seaborg, and others all saw the potential that next-generation reactors have. These reactors can provide clean, affordable, reliable, and extremely safe carbon-free power at a global scale. Some of the key attributes of advanced reactors include:

- Competitive economics due to reduced capital costs and shortened construction times
- Multiple energy output streams ranging from electricity to process heat
- Improved fuel efficiency and the ability to consume used nuclear fuel
- Flexible operations such as load following and grid stabilization to couple with renewables
- Passive and inherently safe designs, producing "walk-away safe" technologies
- Flexible siting, independent of access to cooling water

Additionally, advanced reactors enable a broad diversity of reactor sizes. Micro-reactors, like ours, can bring clean, affordable, and reliable nuclear power to areas that cannot support larger plants. Alaska and Hawai'i are good examples, but there are a number of places in the continental United States, as well as other U.S. territories, that are excellent candidates for these reactors. The size and characteristics of our reactors also enable us to reach markets that are underserved by existing energy technologies. Looking further afield, advanced reactor technologies can fuel mankind's ambitions of navigating the stars. We need energy to explore the heavens, and nuclear energy will power future trips to our neighboring planets and beyond. This is not science fiction. This work is happening today.

The Advanced Reactor Industry

Dozens of startups and large companies are now working to commercialize advanced reactor technologies in the United States. Advanced reactor commercialization efforts have grown significantly in the past decade, particularly in the last five years, and these efforts are better equipped than ever to bring these technologies to market. Nuclear innovation is alive and well, and advances in computational modeling and simulation, along with an injection of talented, creative, and hungry young engineers into the nuclear industry have fueled much of this growth. Federal efforts to attract students into nuclear engineering programs over the last decade are paying dividends, and there is more to come. Furthermore, advanced reactor research and development activities sponsored by the Department of Energy and the national laboratories over the past few decades have demonstrated much of the core technologies that these startups and larger companies are working to commercialize.

This activity has also attracted over $1 billion in private investment. These investors are supporting advanced reactor companies because of their massive market potential, as well as the environmental benefits of next-generation reactors. Many see advanced nuclear as the only practical way to tackle climate change. Some have not invested in energy before, but the compelling business case for advanced nuclear has changed the equation for them. And while the capital invested so far is significant, there is still much more that can and will be invested in advanced nuclear reactor projects, especially if a few opportunities are pursued.

Opportunities and Challenges

Advanced reactor developers face a variety of hurdles and challenges to deploying their reactors. One of the most important is the regulatory process, which is a significant and necessary challenge that advanced reactor developers must navigate. Unfortunately, the regulatory process as it exists today is not a good fit for these new technologies and the venture finance models that fund them. On the other hand, I must emphasize that the widely-held idea that advanced reactors cannot be licensed today is also mistaken. We have found clear licensing pathways for our technology, but at the same time, there is room for significant improvement and modernization.

We support Senate Bill S.2795 which seeks to address many of the issues with the regulatory process today, such as implementing staged licensing processes, risk-informed and performance-based frameworks, reformed hearing schedules, and revised fee structures.

It is also important to mention the progress made by the NRC towards supporting advanced reactor licensing. Recent work with the DOE to develop advanced reactor design criteria, and new guidance on digital instrumentation and control and mechanistic source term quantification will have a substantial effect on advanced reactor licensing.

We also encourage reform on security and staffing requirements so they are "right-sized" to reactor size. Furthermore, future regulatory reforms should yield requirements and cost burdens that reflect reactor size and safety performance. I would also like to acknowledge the work NuScale has done to address many of the challenges that face small reactor developers and the use of modern digital technologies. The work they are doing is paving the way on these issues, from which we will all benefit.

The legacy of R&D by the DOE and its predecessors in the last 70 years has been tremendously helpful to advanced reactor developers today. We are all building on that work. More recently, DOE has supported multiple programs that are helping accelerate advanced reactor development more quickly, but there is still more we can do. The GAIN program and their small business vouchers are a good example of recent efforts, and I would like to see this program continue and expand. We also laud the recent Energy Bill and the support for this kind of work.

DOE's work on advanced reactor fuel characterization and qualification has been and will continue to be quite valuable to advanced reactor commercialization efforts. The facilities and resources used for this work are good examples of just some of the capabilities within the national laboratory complex from which we and other advanced reactor developers can benefit. GAIN provides an avenue for streamlined access to DOE facilities and expertise, and continued initiatives within the GAIN program will help propel advanced reactor efforts. Additionally, DOE sites could be ideal proving grounds for first-of-a-kind reactors. NuScale and INL are paving the way here, as NuScale plans to build their first plant in Idaho; these relationships and processes should continue to be improved and modernized. As we approach the deployment of our earliest reactors, loan guarantee programs will also continue to be crucial for accelerating advanced reactor manufacturing and construction.

Another DOE resource that is often underappreciated is their inventory of nuclear fuels. Demonstration, prototype, and first-of-a-kind advanced reactors will require a variety of fuels, and we would all benefit from being able to use some of the fuel that DOE manages. DOE should anticipate these opportunities, and manage their fuel resources accordingly, maintaining fuels in usable forms and compositions. This may also reduce DOE's fuel management burdens.

The advanced reactor industry has significant global potential. Unfortunately, recent changes to nuclear technology export rules will hinder global growth for U.S. companies. These rules are outdated and need to be modernized so that this growing industry can flourish.

Finally, I must warn DOE (or any other federal agency) of the dangers of playing "kingmaker" or trying to pick winners or losers in the advanced reactor industry. That could severely damage the rising advanced reactor movement. We all have a vector on getting to market and achieving cost parity (or superiority) to coal and gas, while producing carbon-free power at a global scale. Getting there will be hard, but this is one of the most promising times for nuclear since the birth of the industry.

Closing Thoughts

The advanced reactor industry in the United States is growing rapidly. Innovation in nuclear is proceeding at a pace reminiscent of the early days of nuclear power, with dozens of startups and over $1 billion in private capital at work developing the future of energy technologies. The United States is still the global leader in nuclear technology, and we have taken steps to cultivate this growing movement, but there is still more to be done to remove outdated obstacles, and overcome hurdles that slow the growth of this industry. We have a unique opportunity in front of us. If we seize it, we can lead the world in a clean energy transition powered by advanced reactors that can mitigate the effects of climate change, bring affordable, reliable, emission-free energy to the billions without it, and support the growth of an entirely new technology and manufacturing workforce. I thank you for this opportunity to testify, and would be pleased to respond to any questions you might have today or in the future.

Biography

Jacob DeWitte is the co-founder and CEO of Oklo, formerly UPower, a Sunnyvale, CA based company developing an advanced 2 MW reactor. Jacob has been working with nuclear technology since his childhood, and has deep experience with nuclear reactor design and analysis. He has worked with many advanced reactor designs including sodium fast reactors, molten salt reactors, and next-generation PWRs. Jacob has also been involved with front-end and back-end nuclear fuel cycle technology development and analysis. Jacob led the reactor design efforts on a waste consuming molten salt reactor at the University of Florida, and led core design at GE for their PRISM sodium cooled fast reactor. Jacob has also worked at Sandia National Labs, Urenco US, and the naval reactor research laboratories. Jacob is originally from Albuquerque, NM. He completed his undergraduate studies in nuclear engineering at the University of Florida and his SM and PhD in nuclear engineering at MIT.

Figure 1: Rendering of potential site with four reactors

The CHAIRMAN. Thank you, Dr. DeWitte.
Dr. Gilleland, welcome.

STATEMENT OF DR. JOHN GILLELAND, CHIEF TECHNICAL OFFICER, TERRAPOWER

Dr. GILLELAND. Thank you.

My name is John Gilleland. I'm the Chief Technical Officer, as was mentioned. I'm, sort of, CEO Emeritus, I suppose.

It is a nuclear design company based in Bellevue, Washington. I'd like to thank you, Chairman Murkowski and the members of the Committee, for the invitation to testify here today and to extend my particular thanks to our home state Senator, Maria Cantwell, who has been a strong supporter of our operations. Thank you.

TerraPower's goal is to bring our technologies to markets globally as sources of clean, non-emitting, affordable, base load energy and electricity. TerraPower is the developer of the traveling wave reactor, a full size, sodium cooled, fast reactor.

We are also working with Southern Company, Oak Ridge National Lab and the Electric Power Research Institute in the early R & D phase of a very high-potential additional generation for a reactor. We won a DOE-advanced reactor concept award for this activity in January.

Today I would like to talk about the traveling wave reactor because we've come a long way and we'd like to convey the lessons we've learned over the last ten years. We've been at this for ten years.

In 2006 Bill Gates, our Chairman, convened a group of colleagues from the world of science and technology to address two issues, energy poverty and climate change.

Many inhabitants of developing countries have little access to affordable base load electricity. Living standards cannot rise without electricity. Hospitals cannot function without access to reliable power. A child cannot do homework without light to read.

It's been known for a long time that access to electrical energy is essential to human development, but the global consensus of scientists is that climate change requires us to radically reduce carbon emissions. Since developed countries and now developing countries are meeting the population's demands for base load power by burning large amounts of fossil fuels, the resulting emissions are locking us into a deteriorating spiral of climate change and damage to our environment.

Ten years ago Bill Gates and his colleagues looked at the entire menu of low carbon energy solutions. They concluded that nuclear power is an essential element of any credible, low carbon emissions solution. For the right uses and the right venues, wind and solar will play valuable roles but nuclear is the only known technology that can provide the needed huge amounts of energy with minimum impact on our land use and thus, on the natural world.

Nuclear power has already demonstrated its ability to generate large scale, dependable electricity without emissions at affordable prices, and the new nuclear plants now being constructed are setting new standards for accident prevention.

We, at TerraPower, are now ten years and hundreds of millions of dollars into our advanced reactor development. As we've guided those efforts there have been many lessons learned. When I talk to students I usually talk about eight to ten of these lessons. For this Committee, I think there are two main lessons.

First, TerraPower is already using Federal facilities such as the Idaho National Laboratory. Like other companies TerraPower pays to access the government's highly qualified, skilled researchers and advanced equipment. Ours is already an example of public/private partnership.

The bulk of the funds I just mentioned, all from private visionary investors, have gone to universities, businesses and the national laboratories. This is in the spirit of the recent Paris meetings, and we are an early prototype for the breakthrough Energy Coalition's Mission Innovation goals.

Neither the national labs, nor private enterprise could have accomplished what they have done without each other. The recent White House summit on nuclear energy endorsed this approach. The Gateway for Accelerated Innovation in Nuclear, or GAIN, aims to integrate the capabilities of the private sector, universities and laboratories.

So the first big lesson taught to us over the last ten years is that programs like GAIN do work. That's what we've been doing for ten years.

The second big lesson is the government needs to supplement and help the private sector with appropriate and solid oversight functions. We've had good experiences over the last ten years.

One of the effective coordination—one is the effective coordination of TerraPower's international activities with the Department of Energy's National Security, Nuclear Security Administration and the Department of State. Another is the helpful set of consultations we've had with the Nuclear Regulatory Commission (NRC).

But the lesson learned, as we look into the future and we lay out our particular plans, is that it is increasingly clear that Congress must ensure the NRC has sufficient know how and funding to license this country's next generation of nuclear plants.

In closing, I would say our efforts on the TWR and MCFR, which is a Molten Salt Reactor we're doing with Southern Company and others, are but two designs. So we encourage exploration of other innovations such as was talked about by the gentleman to my right.

It is only by working together that we will achieve the breakthroughs we need to make advanced reactors in a better world, a reality. The United States possesses unique strengths, technical and cultural, that can make astonishing accomplishments, if and only if we have the wisdom to unleash them.

Thank you.

[The prepared statement of Dr. Gilleland follows:]

Testimony
John Gilleland, Chief Technical Officer, TerraPower,
Senate Energy Committee, 366 Dirksen SOB, 10:00AM 17 May 2016
THE STATUS OF ADVANCED NUCLEAR TECHNOLOGIES

Good morning. My name is John Gilleland. I am Chief Technical Officer of TerraPower, a nuclear design company based in Bellevue, Washington. I would like to thank Chairman Murkowski and the members of this committee for the invitation to testify here today and to extend my particular thanks to our home state Senator, Maria Cantwell, who is such as a strong supporter of our operations.

TerraPower's goal is to bring our technologies to markets globally as sources of clean, non-emitting, affordable base load electricity. TerraPower is the developer of the Traveling Wave Reactor (TWR), a full size, sodium cooled, Generation IV fast reactor. We are also working with Southern Company, Oak Ridge National Lab and the Electric Power Research Institute in the early R&D phase of an additional Generation IV reactor technology that received a DOE Advanced Reactor Concept award in January.

Today I will talk about the traveling wave reactor because we have come a long way and we would like to convey the lessons we have learned over the last ten years.

In 2006, Bill Gates, our chairman, convened a group of colleagues from the world of science and technology to address two issues: energy poverty and pollution.

Many inhabitants of developing countries have little access to affordable base load electricity. Power outages and load shedding cripple manufacturing and extractive industries. Hospitals cannot function without access to reliable power. A child cannot do homework without light to read. As President Obama recognized in his "Power Africa" initiative, the road to economic and human development is blocked if affordable, dependable power is unavailable. It has been known for a long time that access to electrical energy is essential to human development.

But the global consensus of scientists is that climate change requires us to radically reduce carbon emissions. Since developed countries and now developing countries are meeting the populations' demand for base load power by burning large amounts of fossil fuels, the resulting emissions are locking us into a deteriorating spiral of climate change and damage to our environment.

Bill Gates and his colleagues looked at the entire menu of low carbon energy solutions. They concluded that nuclear power is an essential element of any credible low carbon emissions solution. For the right uses in the right venues, wind and solar can play valuable roles, but nuclear is the only known technology that can provide the needed huge amounts of energy with a minimum impact on our land use and thus on the natural world.

Nuclear power has already demonstrated its ability to generate large scale, dependable electricity without emissions at affordable prices. And the new nuclear plants now being constructed are setting new standards for accident prevention.

But we can use 21st century technologies to do even better ..... much better. Even greater safety improvements, significantly reduced waste production, great extension of fuel supplies, reduction in weapons and terrorist risks and last but not least, lower cost of carbon-free electricity are all possible.

Our flagship technology, the traveling wave reactor, or TWR, offers improvements in all those areas. Its use of a sodium coolant at atmospheric pressure combined with innovative new fuel designs enables operation with far greater safety margins while producing much less waste. It produces only one-fifth of that produced by existing plants. This amount would fill only one and a half rail cars over the plant's 60-year lifetime. The energy value of each pound of mined uranium is increased by more than an order of magnitude and the need for enrichment facilities is greatly reduced.

I would like to conclude with two remarks. The private sector must, and is, stepping forward to develop a new generation of commercially viable plants.

TerraPower is already using federal facilities such as the Idaho National Laboratory. Like other companies, TerraPower pays to access the government's highly qualified, skilled researchers and advanced equipment. Ours is an example of public-private partnership. The bulk of our funds, all from private visionary investors, have gone to universities, businesses and national laboratories. This is in the spirit of the recent Paris meetings and the Breakthrough Energy Coalition's mission innovation goals.

The recent White House summit on nuclear energy endorsed this approach. The "Gateway for Accelerated Innovation in Nuclear" or GAIN, aims to integrate the capabilities of the private sector, universities and laboratories. If Congress provides the labs and universities with the resources, the labs and industry can grow together. We believe this is essential to leverage our strengths and make Gen IV a reality.

Second, the government needs to supplement private sector efforts with a solid oversight function. Already, TerraPower coordinates our international activities regularly with the Department of Energy's National Nuclear Security Administration and the Department of State. Similarly, we consult with the Nuclear Regulatory Commission. We urge Congress to ensure that the NRC has sufficient know-how and funding to license this country's next generation of nuclear plants.

Our efforts on the TWR and the MCFR are two designs. We encourage exploration of other innovations as well. It is only by working together that we will achieve the breakthroughs we need to make advanced reactors and a better world a reality.

Thank you for your time.

The CHAIRMAN. Thank you, Dr. Gilleland.
Mr. Hopkins, welcome.

STATEMENT OF JOHN HOPKINS, CHAIRMAN AND CHIEF
EXECUTIVE OFFICER, NUSCALE POWER

Mr. HOPKINS. Thank you, Senator.

NuScale Power is currently the leading developer of American small modular reactor technology. This technology has been in development for more than 15 years. Our company is based in Corvallis, Oregon and majority owned by the Fluor Corporation.

We are advancing a unique and innovative SMR design which offers the safest light water reactor nuclear technology that is near term deployable. Our design is uniquely safe. We have solved one of the most vexing problems of the nuclear industry with what we call the triple crown of nuclear safety.

In a case of a loss of all sources of electricity at the plant, the NuScale Power module shuts itself down and cools for an unlimited period of time. With no operator action required, no need for additional water and no electricity.

The NuScale Power module uses simple properties of physics, convection, conduction and gravity to drive the flow of coolant in the reactor. The thermal hydraulic properties and capabilities of technology have been demonstrated through an extensive test program inspected by the U.S. Nuclear Regulatory Commission and which are protected by patents issued or pending since 2011.

The NuScale Power module is an ideal option for carbon-free electricity generation. The NuScale design is dramatically smaller than today's pressurized water reactors and eliminates need for safety-related, electrically-driven pumps, motors and valves necessary to protect the nuclear core. It can be factory-manufactured and transported to a site via rail, truck or barge.

We are preparing for our first deployment project, known as the Utah Associated Municipal Power Systems Carbon Free Power Project, which will be sited in Idaho and possibly a location on the Department of Energy's Idaho National Laboratory site.

We expect to deliver our first project of 12 modules in a 600 megawatt plant to UAMPS for an overnight price of less than $3 billion with commercial operation commencing in 2024.

Energy Northwest which operates the Columbia Generating Station in Washington State, has joined this project and holds a first right of offer to operate the UAMPS project.

In December 2013 the Department of Energy selected NuScale as the sole awardee for funding in round two of the DOE Small Modular Reactor Licensing Technical Support Program focusing on providing a cost share grant in support of licensing expenses. NuScale may receive up to $217 million of matching funds over five years. We are the only near-term deployable SMR developer receiving DOE funding support, and we are proceeding at full speed toward long-term commercialization.

With support from this funding NuScale has expanded its workforce to include more than 600 engineers and has made substantial progress on the engineer and analysis and tested needed to complete the design certification application for submittal to the NRC by the end of 2016. Successful completion of the DOE-funded SMR cost share program depends on sustained congressional support through continued appropriations. We appreciate your past support, and we ask that you continue to prioritize small modular reactor programs in a tight budgetary environment.

A risk to the delivery of our technology as currently planned is the uncertainty of the time and process for the NRC design certification and combined operating license efforts. In order to meet our customer needs to deliver carbon free electricity to their grids, we must be positioned for commercial operations in 2024.

NuScale has been engaged with the NRC in pre-application review efforts since April 2008. We are on schedule to submit our design certification application by the end of this year, and the NRC plan currently reflects a 40-month review process. We are currently working with senior staff at NRC to complete the final issuance of the new scale design specific review standard expected by the end of June 2016 which provides the acceptance criteria for their review of our DCA.

We appreciate the quality interactions we continue to have with the Office of New Reactors and their dedication to a thorough and timely review. It is important that sufficient NRC resources are assigned to review in a NuScale application to ensure completion within the 40-month schedule so that we can be in position to meet the growing marketplace demands for our carbon free energy source.

I'd like to thank the Committee for holding this important hearing. And I look forward to any questions you may have.

Thank you.

[The prepared statement of Mr. Hopkins follows:]

Testimony of NuScale Power before the
Committee on Energy and Natural Resources of the
U.S. Senate

Hearing to Examine the Status of Innovative Technologies Within the Nuclear Industry

Testimony provided by John L. Hopkins, Chairman and Chief Executive Officer, NuScale Power

May 17, 2016

**Testimony of NuScale Power before the
Committee on Energy and Natural Resources of the
U.S. Senate**

Hearing to Examine the Status of Innovative Technologies Within the Nuclear Industry

Written Testimony provided by John L. Hopkins, Chief Executive Officer, NuScale Power

May 27, 2016

NuScale Power is the leading developer of American Small Modular Reactor (SMR) Technology. For more than 15 years, our innovative company, based in Corvallis, Oregon and majority-owned by the Fluor Corporation, has been advancing a unique SMR design which offers the safest nuclear technology which is deployable in the near term. This significant advance, coupled with the deployment characteristics of our SMR design, can play a significant role in the future needs for baseload carbon-free electricity generation.

The genesis of our 50 MWe integral pressurized water reactor began over 15 years ago with a U.S. Department of Energy (DOE) grant through the Idaho National Laboratory and included the construction of a one-third scale electrically-heated prototype test facility to validate the safety features of the plant. This prototype has been in operation since 2003.

Unique Safety Features

First, I will speak about the safety features of the NuScale SMR plant design. We eliminated many of the complex systems found in existing operating nuclear power plants and replaced them with a design which emphasizes the use of the natural forces of physics. The result of this unique design is a nuclear plant immune to the effects of a complete loss of all electricity sources to the generating facility, such as the situation that was experienced at Fukushima.

NuScale Announces Major Breakthrough in Safety
Wall Street Journal April 16, 2013

- NuScale design has achieved the "Triple Crown" for nuclear plant safety. The plant can safely shut-down and self-cool, indefinitely, with:

  - **No Operator Action**
  - **No AC or DC Power**
  - **No Additional Water**

- Safety valves align in their safest configuration on loss of all plant power.

- Details of the Alternate System Fail-safe concept were presented to the NRC in December 2012.

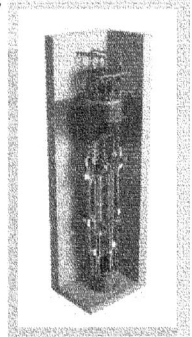

Nonproprietary
©2014 NuScale Power, LLC

NuScale POWER

As shown in this illustration, we have solved one of the most vexing problems of the nuclear industry with what we call the Triple Crown of Nuclear Safety. In the case of an event where all sources of electricity are absent, the NuScale Power Module (NPM) shuts itself down and self-cools for an indefinite period of time, with no operator action required, no need for additional water other than the 7.4-million gallon pool the NPMs are already immersed in, and no electricity. This capability has been demonstrated and witnessed by the U.S. Nuclear Regulatory Commission (NRC) and is protected by patents issued or pending since 2011.

How exactly does this work?

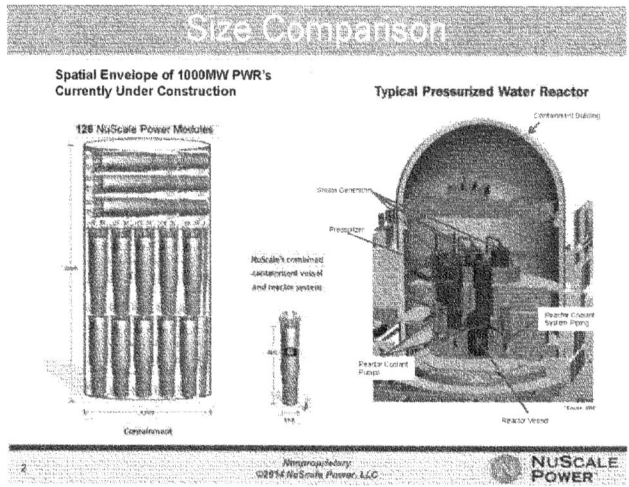

This picture illustrates the size of a 50MWe NPM compared to a typical 1000MWe-class plant operating today (on the right) and the spatial envelope of the containment buildings being constructed today in Georgia and South Carolina (on the left). The image on the right shows reactor coolant pump motors, steam generators, a reactor containment building, pressurizer and hundreds of feet of large-diameter thick-wall reactor coolant system piping through which approximately 20 million gallons per hour of high temperature reactor coolant water flow. The center image is the NPM which is a complete integrated nuclear steam supply system. It is a factory-built unit, including the containment, reactor vessel, steam generators and pressurizer, all contained in one cylindrical, road-transportable vessel (Note: By design the NPM does not contain reactor coolant pumps nor large diameter reactor coolant system piping).

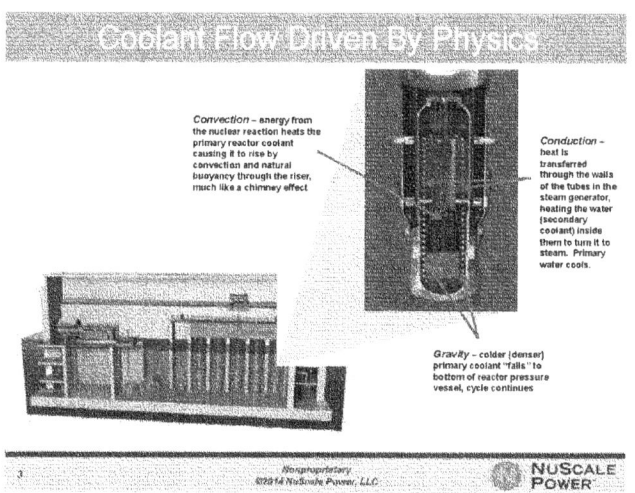

The NPM uses simple forces of physics to drive the coolant flow, as shown in this illustration. The image on the right shows the NPM which includes an outer steel vessel (containment vessel) containing an inner vessel (reactor vessel), installed underground and in a 7.4-million-gallon pool of water. NPMs are installed in a reactor building pool designed to accommodate up to 12 NPMs. The NPM operates when the heat from nuclear fission in the reactor core, represented by the red area at the bottom of the diagram, heats pressurized water causing the water to rise by buoyancy and convection through the bronze-colored tube, much like a chimney-effect. The bronze-colored tube is surrounded by coiled tubes containing cooler water. As the hot water passes over these tubes it gives up energy by transferring heat from the hotter water to the cooler water by conduction through the walls of the tubes, causing the internal water to turn to steam. The steam is then directed to rotate a turbine and generate electricity. As the hot water gives up energy, it becomes cooler, and thus denser, which causes it to fall by gravity to the bottom of the reactor vessel where the natural circulation flow cycle continues. The NPM eliminates many of the electrically-driven components required to protect the core found in today's conventional nuclear plants. That is, in the NuScale design, electricity is simply not required to protect the core.

The Role of the U.S. DOE in SMR Technology Development

In the early 2000's, Congress authorized a program known as NP-2010, to stimulate the revival of the U.S. nuclear industry with a cost-share of the private sector investments in design and licensing. This program resulted in the certification of two new nuclear plant designs, one of which is being built today in Georgia and South Carolina. Of note, the design certification testing for that design was performed under contract to the designer by NuScale founder and Chief Technology Officer, Dr. Jose Reyes, on facilities he designed and which were constructed adjacent to where the NuScale test facility is located in Corvallis, Oregon.

Congress also authorized a similar program for Small Modular Reactor design and licensing cost sharing. In December 2013 DOE selected NuScale as the sole awardee for funding in round two of DOE's SMR program, focusing on providing cost-share grants in support of licensing expenses. As such, we may receive up to $217 million of matching funds to aid in funding the approximately $1 billion necessary to complete the design and license it for construction. We have spent approximately $420 million on the project to-date, including about 60% of our DOE award. Successful completion of the SMR cost-share program depends on sustained Congressional support through continued appropriations. We appreciate your past support, and ask that you continue to prioritize SMR programs in a tight budgetary environment.

A substantial portion of DOE's cost-shared funding will help pay for NRC fees required of all NRC applicants. To date, NuScale has incurred over $6 million in NRC fees to pay for the approximately 20,000 hours of NRC staff time utilized to date. We estimate we will incur a total of approximately $60 million in NRC fees by the time we complete the 40-month review of our design application.

The Importance of Timely NRC Design Certification and Combined Operating License Actions

In order to ensure that the NuScale design is ready to meet the needs of our prospective customer base as they begin to reduce their reliance on carbon-generating assets, we will need the NRC to complete preparations for the receipt of our design certification application and to conduct the design certification review in a timeframe that meets these needs. We hear two consistent concerns from prospective customers and investors: 1) regulatory uncertainty—will regulations change in mid-process?, and 2) skepticism the NRC can complete their review in the 40 months the NRC has planned.

NuScale is aggressively working with the NRC to address both of these concerns. With respect to regulatory uncertainty, we have engaged the NRC in pre-application submittal preparation since early 2008, in order to prepare the NRC for how we will develop the design certification application in the context of the unique features of our design. Together with the NRC we have collaboratively identified additional remaining technical, licensing and policy issues to be addressed by the end of 2016. With respect to the NRC's ability to conduct the Design Certification Application review within 40 months, we see two critical issues to address: 1) extending the NRC's review time on critical areas by submitting portions of the design in advance; and 2) obtaining dedicated NRC staff resources to review our design. To address the first issue NuScale is submitting portions of our application early, in the form of what are known as topical reports, on 13 specific aspects of the design. Seven of these are under NRC review and six will be submitted in the next 4 months. By submitting these reports, we effectively extend the staff's review time and reduce the scope of work that needs to be done when the remainder of the application is submitted. Regarding the second issue, the NRC reviewed the initial wave of applications submitted for design certifications using a matrix organization, where the same staff person reviewed multiple designs. While that approach is efficient and makes sense when all the designs are similar, such an approach may create more schedule risk for a unique design such as NuScale's. Therefore, we believe a dedicated NRC staff review team, whose members' sole function is to review the NuScale design, is more likely to meet the NRC's commitment to 40 months. Such a focused approach would help to ensure that sufficient resources are identified for the review in advance of the submittal and that staff can be trained on the unique safety features of our design prior to submittal of the application. We have been working collaboratively with the NRC to help develop this approach.

Customers and Markets

The most important element for the success of our program is whether there will be market demand for non-carbon-generating baseload electricity. Our prospective customers are faced with increasing carbon regulation, accelerated coal plant retirements, and increased integration of intermittent renewable generation assets from wind and solar. The NPM is uniquely positioned to provide a baseload resource that meets these needs and is complementary to renewables by being designed for load-following. And, since a NuScale plant consists of twelve individual power modules, it has the ability to load follow incrementally by varying the output of individual modules to match variations in intermittent generation.

The first deployment of a NuScale plant is currently designated for the Utah Associated Municipal Power Systems (UAMPS) for the project known as the "UAMPS Carbon Free Power Project (UAMPS CFPP)." UAMPS's project contemplates commercial operation of the first NPM in 2024. The plant will be located in Idaho, possibly at the DOE's Idaho National Laboratory site. The exact location is currently being identified through a rigorous site selection process. Energy Northwest has joined this project by holding the first right of offer to operate the UAMPS project.

The NuScale plant is expected to be delivered to UAMPS for a price of approximately $3 billion. We plan to construct it on a schedule of approximately 36 months from the start of safety-related construction through commissioning of the first module.

Conclusion

NuScale is proud to be developing a new innovative Small Modular Reactor technology for the deployment of a nuclear power generating asset that sets a new safety standard for nuclear energy facilities, is carbon-free, factory-built, incrementally deployable, and significantly less costly than large generating units. The NuScale Power Module is a disruptive technology that will change the way the world views nuclear energy, and it will play an important role in next generation deployment of baseload electricity. We are grateful for the support of the U.S. DOE and the Congress. We take the responsibilities associated with the use of taxpayer dollars, very seriously and are singularly focused on our role in the re-establishment of U.S. leadership in small modular reactor nuclear technology.

The CHAIRMAN. Thank you, Mr. Hopkins.
Mr. Kuczynski, welcome.

STATEMENT OF STEPHEN KUCZYNSKI, CHAIRMAN, PRESIDENT AND CHIEF EXECUTIVE OFFICER, SOUTHERN NUCLEAR OPERATING COMPANY, INC.

Mr. KUCZYNSKI. Welcome.

Good morning, Chairman Murkowski, Ranking Member Cantwell and members of the Committee. My name is Steven Kuczynski. I'm the Chairman, President, CEO of Southern Nuclear Operating Company. We operate a fleet of six nuclear power reactors at three sites and are constructing two state of the art AP1000. That's Advance Pass of 1000 reactors at Plant Vogel near Augusta, Georgia. It's an honor to appear before this Committee to share my views on advanced nuclear technologies, an area that is pivotal to our nation's future and worthy of this Committee's interest.

I currently serve as Chairman of the Nuclear Energy Institute Advanced Reactor Working Group, or ARWG, an industry initiative created with the understanding that today's decisions about nuclear energy, research and development will enable a nuclear fleet of the future. Today the U.S. fleet of 100 commercial reactors which provides 20 percent of the nation's electricity is comprised exclusively of light water reactors. They are called light water reactors because they use ordinary water for cooling purposes and to moderate the nuclear chain reaction.

In some other countries, such as Canada, deuterium oxide is the main coolant to moderator. Deuterium oxide is heavier than water which is why these reactors are commonly known as heavy water reactors.

Taken together light water and heavy water reactors have proven to be the safest, most efficient, cost effective means of electricity generation.

Within the ARWG we see tremendous promise for even better, more innovative nuclear reactors, and we are actively working toward achieving demonstration of multiple advanced reactors by 2025. We believe commercial deployment is feasible by 2035.

Advanced reactors are often called non-light water reactors because they do not use water as a coolant or a moderator but neither do they use heavy water. Instead, advanced reactors under consideration in the United States are being designed around the use of liquid metals, salts, gases or other advanced techniques.

At Southern Company we are actively engaged in researching and assisting in deployment of advanced reactor technologies including the prismatic block high temperature gas cooler reactor which is expected to be significantly more efficient than current operating reactors and the molten chloride fast reactor which we are exploring under the DOE Advanced Reactor Concepts Program alongside TerraPower, Oak Ridge National Laboratory, the Electric Power Research Institute and Vanderbilt University. We are very proud to be working with these distinguished organizations.

So one might ask, why support advanced reactor research and development?

We know that significant new electric generating capacity will be required in the decades ahead to meet the nation's growing energy needs. Nuclear power is an attractive option for meeting this demand with reliable, affordable, clean sources of base load electricity with zero emissions. And advanced reactors will be even more efficient, produce less by product material, have enhanced safety features, require even a smaller geographic footprint, come at a lower cost to customers and be capable of using a broad range of fuel types.

In addition, advanced reactors are expected to be scalable, meaning they can be constructed in varying sizes, and able to adjust output to meet variable demands or supplement intermittent renewables. They will also provide a valuable source of process heat for a wide range of industrial customers, and we see tremendous promise for military installations to use these technologies to generate electricity and provide energy to meet other needs.

In addition to the work we're doing with Gen IV reactors, Southern Nuclear is engaged with the industry's effort to bring small modular reactors to market.

In closing I'd like to highlight five key points for the Committee.

First, collaboration between the Federal Government and the private sector will be critical to promoting nuclear energy innovation in the United States. As was true in the early days of nuclear technology development, the national labs will need to play a central role. Importantly, the national labs have the resources and facilities, as well as the flexibility within the existing regulatory structure to accomplish significantly more advanced R and D work than a private U.S. company or university could do alone.

Second, the Federal Government needs to move forward with innovative licensing frameworks and regulatory structure tailored for advanced reactors. The current regulatory system with its exemption based licensing approach built around light water reactors will be ineffective for licensing non-light water technologies, makes it difficult to attract private investment. Southern Company is taking a lead role in industry led licensing modernization initiatives.

Third, innovation benefits from competition which is why the Federal Government should support advanced reactor programs without picking the ultimate winners and losers at this point.

Fourth, the U.S. must remain the global leader in nuclear energy technology, as we have been in the past. Many nations are currently working on advanced reactor designs. Our nation should not cede leadership in nuclear power innovation to others.

Finally, the ARWG's goal of commercial deployment by 2035 is, in my view, achievable. With the current fleet of reactors, it took just a decade to progress from a concept to commercial operation. This required innovation, collaboration, leadership from forward thinkers like Admiral Rickover and the strong support of Congress and the Executive Branch.

Our nation has the knowledge and expertise to make this kind of technological progress in nuclear energy happen again. As before, innovation and collaboration will provide the keys to success. Again, thank you for allowing me to appear before your Committee, and I look forward to your questions.

[The prepared statement of Mr. Kuczynski follows:]

Statement of Stephen E. Kuczynski

Chairman, President and Chief Executive Officer

Southern Nuclear Operating Company, Inc.

BEFORE THE U.S. SENATE COMMITTEE ON ENERGY & NATURAL RESOURCES

"Hearing to Examine the Status of Advanced Nuclear Technologies"

May 17, 2016

EXECUTIVE SUMMARY

> Nuclear power is a leading source of affordable, reliable, clean, American energy that powers our economy, protects our national security, preserves the environment, provides high-paying jobs for thousands of our fellow citizens, and generates substantial tax revenue for local, state, and federal governments. Our nation's current fleet of nuclear reactors is aging and will eventually need to be replaced with new reactors in order to ensure that nuclear power—an emission-free source of electricity—will continue to provide a significant share of America's electricity needs. New reactors also provide a valuable option for replacing other retiring electric generating units and for meeting energy demand growth.

> Southern Company is proud to lead the nation by constructing two first-of-a-kind, "Generation III+" nuclear units at the Vogtle Electric Generating Plant near Augusta, Georgia. Taken together, these state-of-the-art Westinghouse AP1000 units are projected to supply over 2,200 megawatts (MW) of new, baseload, zero-emission electric generation, creating more than 5,000 total construction jobs and over 800 permanent jobs.

> Even as we make significant progress toward commercial deployment of Generation III+ reactors, we are already exploring the next generation of nuclear technologies known as "Generation IV" (or "Gen-IV") reactors. On January 15, 2016, the Department of Energy (DOE) selected a Southern Company-led proposal as one of two recipients of approximately $6 million for this year (up to $40 million over the next five years) to explore, develop, and demonstrate advanced nuclear technologies. With non-federal cost-share contributions, this project represents up to $80 million in new advanced reactor research. Our partners in this public-private partnership are TerraPower, Oak Ridge National Laboratory (ORNL), the Electric Power Research Institute, and Vanderbilt University. We believe Gen-IV advanced reactors will build on the Gen-III+ advantages, with even more advanced safety systems, less byproduct materials, and greater cost efficiencies. Advanced reactors will also serve as a source of process heat for various industrial applications.

> Innovation and technology are engines of American greatness. Demonstration of advanced reactors by 2025 and commercial deployment in the 2030-2035 timeframe are ambitious yet achievable goals. This mission will require public-private collaboration, resulting in innovative policies, licensing frameworks, and regulatory structures that facilitate the efficient and predictable deployment of these new technologies and encourage private investment. As a range of technology options is explored, we will advocate for and encourage industry-led collaboration with DOE, vendors, utilities, universities and national labs to leverage capabilities and share some of the risks.

> Bringing advanced reactor technology to market will take a level of effort and commitment well beyond the status quo but the benefits to the American public, in terms of national security, global leadership and economic competitiveness, development of high paying jobs, and the environment, are immense and justify this Committee's close consideration of steps that the federal government can take to support these endeavors.

Good morning Chairman Murkowski, Ranking Member Cantwell, and Members of the Committee. Thank you for the opportunity to appear before you today.

My name is Steve Kuczynski, and I am the Chairman, President and CEO of Southern Nuclear Operating Company, Inc., where I am responsible for the operation of a fleet of six nuclear power reactors at three sites as well as the construction of two new reactors at Plant Vogtle near Augusta, Georgia. It is an honor to appear before this Committee to share my views on nuclear energy innovation and advanced nuclear technologies, an area that is pivotal to our nation's future and worthy of this Committee's interest and attention. I currently serve as chairman of the Advanced Reactor Working Group (ARWG), an initiative of the Nuclear Energy Institute (NEI) charged with developing an industry vision and execution strategy for a long-term sustainable program that will result in the development and commercialization of advanced reactors. An ambitious goal of the ARWG is to achieve demonstration of multiple advanced reactors by 2025 and commercial deployment in the 2030-2035 timeframe.

During my career, I have been responsible for a wide range of disciplines at nuclear power plants—from safety, training and emergency preparedness to radiation protection, operations, and construction. In my testimony today, I will discuss Southern Nuclear's fleet of nuclear power plants, including the ongoing construction of our two newest reactors. I will also share my personal perspectives on advanced nuclear reactors and the merits of continued governmental and private sector interest and investment. While other co-panelists represent the views of the entities designing and selling advanced reactors, my vantage point is that of an ultimate end-user of these technologies. I appreciate that the Committee has sought to hear from both views this morning.

Southern Nuclear

Headquartered in Birmingham, Alabama, Southern Nuclear is a subsidiary of Southern Company, the nation's premier energy company serving the Southeastern United States through its subsidiaries. Southern Company is a leading U.S. producer of clean, safe, reliable and affordable electricity. Southern Nuclear currently operates six nuclear reactors: Units 1 and 2 at Plant Farley near Dothan, Alabama; Units 1 and 2 at Plant Hatch near Baxley, Georgia; and Units 1 and 2 at Plant Vogtle near Augusta, Georgia.[1] We have been in the nuclear power business for almost 50 years, dating back to Southern Company's decision in 1967 to build Plant Hatch, our very first nuclear power plant, which began commercial operation in 1975. Together, Plants Farley, Hatch and Vogtle provide approximately 20% of the electricity used in Alabama and Georgia. This is made possible by our talented and committed workforce of over 4,000 men and women working at our power plants and corporate offices, who are also part of the larger Southern Company team of over 26,000 employees across the States of Alabama, Florida, Georgia, and Mississippi.

---

[1] Plant Farley is owned by Alabama Power Company. Plants Hatch and Vogtle are co-owned by Georgia Power Company, Oglethorpe Power Corporation, the Municipal Electric Authority of Georgia, and Dalton Utilities.

Nuclear power is a leading source of affordable, reliable, clean, American energy that powers our economy, protects our national security, preserves the environment, provides high-paying jobs for thousands of our fellow citizens, and generates substantial tax revenue for local, state, and federal governments. As a proud corporate citizen of the communities where we operate, Southern Nuclear's top priority is the safety and health of the public and our employees. We are committed to the safe operation of our nuclear generating facilities with equipment and systems that meet rigorous safety and design regulations. Plants Farley, Hatch and Vogtle are national leaders in safe operation and reliability with an average three-year fleet capability factor of 92.4% from 2013 to 2015, which exceeded the national average of 90.3% for the same period.[2] The U.S. Nuclear Regulatory Commission's (NRC) annual assessment for 2015 concluded that our nuclear power plants met all of the NRC safety and security performance objectives. With increasing focus on reducing emissions of carbon dioxide (CO_2), we are proud that our existing fleet of nuclear reactors prevents more than 56 million metric tons of CO_2 from entering the atmosphere, which is the equivalent of taking 10 million cars off the road—more than the number of cars registered in Alabama and Georgia, combined.

We are in the business of generating safe, clean, reliable and affordable energy through a relentless focus on safety, a culture of continuous improvement and innovation, and a commitment to providing value to our customers. Innovation is a central part of our strategy to achieve these objectives. Innovation and technology are engines of American greatness. It is within this context that Southern Company is investing in advanced reactor technology research and development and looking ahead toward the steps needed to promote the licensing, construction and utilization of these technologies.

Delivering the Current Generation of Nuclear Power

Southern Company supports an "all of the above" energy policy and strategy that balances the goals of clean, safe, reliable, and affordable energy and provides a full role for renewables, energy efficiency, new nuclear, 21st century coal, and natural gas. This approach necessitates strong investment in the future of nuclear power.

Nuclear energy provides approximately 20% of our nation's electricity needs and represents over 60% of our nation's emission-free generation. In 2014, nuclear energy facilities prevented 594 million metric tons of carbon dioxide emissions, the equivalent of taking 135 million cars off the road. Furthermore, the current nuclear fleet provides a substantial economic benefit to the country. The average nuclear power plant pays approximately $16 million in local and state taxes and $67 million in federal taxes a year, and provides thousands of high-quality, permanent jobs in their communities and significantly impact the local economies.

---

[2] *See* Fourth Quarter 2015 Data File, World Ass'n of Nuclear Operators (on file with Southern Nuclear Operating Company, Inc.). Capability factor measures the amount of time the plant is on-line and producing electricity. For more information about the nuclear industry's 2015 performance measures, please visit http://www.nei.org/News-Media/Media-Room/News-Releases/Nuclear-Power-Plants-Set-Records-for-Safety,-Opera.

Wind and solar generation are making important contributions to the nation and will continue to do so, but nuclear power retains qualities that other emission-free generation sources do not possess. Nuclear energy's high capacity factor makes it perfectly suited for baseload power—the generation required to meet the minimum level of constant, continuous electricity demanded by electricity customers. Adequate baseload power helps to ensure grid stability, voltage control, and other features essential to reliably powering our economy. Renewables are intermittent sources of power that cannot provide the 24/7, baseload power that nuclear energy has consistently provided for decades. In addition, in order to generate an amount of electricity equal to that of a nuclear reactor, solar and wind plants demand a much larger footprint. Consider these facts: a 1,000-megawatt (MW) nuclear power plant requires just over one square-mile of land; to generate the same amount of power, a wind facility requires anywhere from 260 to 360 square-miles and a solar facility requires 45-75 square-miles.[3]

However, the current fleet of nuclear reactors is aging and will inevitably need to be replaced with new reactors in order for nuclear power to continue to provide a significant share of the nation's electricity needs and a majority of its emission-free generation. According to the Energy Information Administration, the nation will need over 285 GW of new electricity capacity by 2040. Nuclear reactors remain the best, cleanest, safest, and most reliable form of baseload electric generation to meet these energy needs. Additionally, the U.S. Environmental Protection Agency (EPA) has repeatedly acknowledged nuclear power's important contribution to achieving environmental goals.

Congress, too, has consistently endorsed a central role for nuclear power in our energy policies. The Energy Policy Act of 2005 and the Energy Independence and Security Act of 2007 sought to expand the commercial utilization of nuclear energy in the United States, while also reducing CO_2 emissions and ensuring affordable, reliable, and clean domestic energy for Americans. These laws were a catalyst for the construction of new reactors like those at Plant Vogtle Units 3 and 4. Federal policies have often wisely centered on incentivizing nuclear power and ensuring an efficient regulatory approval process. This has included essential loan guarantee programs, licensing reforms, and a host of other measures. Our state partners also recognize the pivotal role that nuclear power should play in our energy future.

Southern Company is proud to lead the nation by constructing first of a kind nuclear units at Plant Vogtle. Taken together, these state-of-the-art Westinghouse AP1000 units are projected to supply over 2,200 MW of new, baseload, zero-emission electric generation, creating more than 5,000 total construction jobs and 800 permanent jobs. These are some of the first new nuclear units to be built in the United States in over 30 years. The Vogtle site is among the largest ongoing construction projects in the United States. This is a joint effort with our power plant ownership team, which includes Georgia Power Company, Oglethorpe Power Corporation, Municipal Electric Authority of Georgia, and Dalton Utilities, and a construction team consisting of Westinghouse Electric Company LLC and WECTEC Global Project Services Inc. (formerly CBI Stone & Webster, Inc.). Throughout the duration of this construction project, just as with the

---

[3] Nuclear Energy Institute, "Land Needs for Wind, Solar Dwarf Nuclear Plant's Footprint," (July 9, 2015), available at http://www.nei.org/News-Media/News/News-Archives/Nuclear-Power-Plants-Are-Compact,-Efficient-and-Re

operation of our existing plants, safety always comes first. We remain focused on completing Vogtle Units 3 and 4 with safety, quality, and compliant construction as top priorities. We will not compromise. Similarly, South Carolina Electric & Gas Company is also demonstrating national leadership by constructing two AP1000 units at the Summer Nuclear Station.

Nuclear energy already has tremendous advantages over other forms of electric generation: zero emissions, capacity factors exceeding 90%, safety records that exceed those of other energy sources, as well as affordability over the long term without the price swings common to other fuels. The AP1000 design adds even more layers of safety redundancies and with a simplified plant design, the AP1000 is less expensive to build, operate and maintain. This includes substantially fewer valves, pumps, piping, building volume, and cable compared to earlier-generation nuclear plants.

Delivering the Next Generation of Nuclear Power

Even as we move toward commercial deployment of "Generation III+" reactors like the AP1000 at Plant Vogtle, we are already exploring the next generation of nuclear technologies known as "Generation IV" (or "Gen-IV") reactors. On January 15, 2016, DOE selected a Southern Company-led proposal as one of two recipients of approximately $6 million for this year (up to $40 million over the next five years) to explore, develop, and demonstrate advanced nuclear reactor technologies. With non-federal cost-share contributions, this project represents up to $80 million in new advanced reactor research. Our partners in this public-private partnership are TerraPower, Oak Ridge National Laboratory (ORNL), the Electric Power Research Institute, and Vanderbilt University. This project will bolster the development of molten chloride fast reactors (MCFR), an advanced concept for nuclear generation under development by TerraPower. As a company, we are proud to be afforded this opportunity and we look forward to seeing additional collaboration to strengthen this partnership through delivering results for our partners, DOE, and the American taxpayer. In addition to the MCFR, we are also assisting in the development of modern Prismatic Block High Temperature Gas Cooled Reactor (HTGR) technologies, which are expected to be significantly more efficient than current operating reactors. Lead-Cooled Fast Reactors, Sodium-Cooled Fast Reactors, Supercritical Water-Cooled Reactors, and other kinds of non-light water reactor technologies are being considered in the United States or abroad.

While Southern Company has not made any commitments toward construction of power plants with MCFR, HTGR, or other advanced reactor technologies, the potential for Gen-IV reactors is enormous. We believe these reactors will build on the Gen-III+ advantages, with more safety systems, less byproduct materials, and greater cost efficiencies. Many of the anticipated advantages to Gen-IV reactors are summarized below in Table 1.

Table 1. Summary of Potential Advantages to Gen-IV Advanced Reactors.

Advantages	Description
More Efficient	Designs are expected to generate more than 100 times the energy yield of current reactors using the same amount of fuel. Operation at high temperatures allows for the generation of process heat to support other industrial operations.
Less Byproduct	Numerous designs have the ability to consume existing used fuel, thereby reducing storage capacity needs.
Zero Emissions	Like existing reactors, Gen-IV reactors will produce no air emissions.
Enhanced Safety Features	While the nation's existing fleet of nuclear reactors is operating safely today, Gen-IV reactors will have increased safety features that far exceed NRC requirements for today's reactors. In fact, with some designs, there is a potential for reduced, or even eliminated, emergency planning zones. Many operate at low pressure and utilize fuels that cannot melt. Simpler designs require fewer components and are less prone to failure. Passive core protection functions can cool the reactor for days at a time without the need for off-site power. Non-water coolants reduce the risk of a loss of coolant accident.
Smaller Footprint	With fewer components, advanced designs take up even less land than the already compact designs of current reactors.
Lower Cost	Simpler designs allow faster, lower cost of construction. Because there is less waste, storage costs are reduced dramatically.
Fuel Diversity	Capable of using a broader range of fuel types including raw fuels that may not require an expensive enrichment process.
Scalable	Can be constructed in varying sizes from diesel generator replacement size to larger than those in the current nuclear fleet. Can adjust output to meet variable demands or supplement intermittent renewables.
Global Leadership & Competitiveness	Gen-IV reactors will further demonstrate the leadership position of the United States in advanced nuclear energy technologies and will enhance our nation's competitiveness in the global economy.

As this information demonstrates, Gen IV reactor technologies build on the advantages of the nation's existing nuclear reactor fleet and provide benefits that make it essential for us to evolve these technologies.[4] With these advantages and features, the U.S. military would also likely find advanced reactor technologies to be promising especially as the nation explores ways to equip and power a lighter, more mobile fighting force. Moreover, advanced reactors will also serve as a source of process heat for various industrial applications such as desalinization, oil refining, and hydrogen production.

In addition to the work we are doing with Gen-IV reactors, Southern Nuclear is also engaged with the industry's efforts to bring small modular reactors (SMRs) to market. In January of this year, Southern Nuclear and several other leading developers and potential customers announced a memorandum of understanding establishing a consortium called "SMR Start," which is designed to help accelerate SMR commercialization.

Innovation Requires Collaboration

As all of these potential technologies are explored, the federal government should support advanced reactor programs without picking the ultimate winners or losers. Innovation requires competition. Within our own company, we take great pride in our culture of innovation and desire for step-up performance improvement in all facets of our business. We also believe that our federal government partners have the capability to create the right environment for innovation in the nuclear technology arena to flourish. This includes public-private partnerships that can harness the power of collaboration.

A shining example of productive collaboration leading to continuous state-of-knowledge improvement is flying above us every day. The International Space Station (ISS) is a truly international scientific and technological collaboration. It orbits the Earth once every 90 minutes, seeing 16 sunrises and sunsets daily. The ISS has been continuously inhabited since 2000 by crews of three to six people from 15 different countries, involving collaboration of five different space agencies representing the United States, Canada, Russia, Japan, and Europe, with the European Space Agency funding coming from 11 European countries.

In much the same way, we cannot achieve sustainability in innovation by ourselves. Collaboration at the private sector, governmental, academic, and international levels will be key to achieving demonstration of advanced reactors by 2025 and commercial deployment in the 2030-2035 timeframe, an effort that will entail creation of innovative policies, licensing frameworks, and regulatory structures that facilitate the efficient and predictable deployment of these new technologies and encourage private investment. I believe it will also require our federal partners to share the cost of state-of-the-knowledge improvements. DOE, universities, vendors and our centers of knowledge will need to leverage the best talent our nation has to offer.

---

[4] For a more comprehensive list of North American Advanced Reactor Projects, please see the June 15, 2015 report, *The Advanced Nuclear Industry*, prepared by Third Way, available at www.thirdway.org/report/the-advanced-nuclear-industry.

Likewise, public-private partnerships are, in the context of advanced reactors, uniquely necessary as these technologies are subject to an extensive regulatory regime requiring complex technical work to assure regulators and the public of the safety of these new reactors. These endeavors also require new fuel types to be developed and tested, and supply chains for new kinds of equipment to evolve and mature. In addition, it will be necessary to design and test prototypes and, ultimately, a first of a kind commercial reactor will have to be designed, approved, constructed and operated. We are already seeing increased private sector investment in proposed new reactor startups and systems reaching, by some estimates, more than $1 billion.[5] Nonetheless, because of the expense, regulatory uncertainty and timeframes involved, continued public sector investment will be necessary to make the leap from the laboratory to commercial deployment.

Additionally, as was true in the early days of nuclear technology development, we must work with our national labs to safeguard the nation's significant investments in nuclear technology, ensure we remain the world leader in this area, and demonstrate newer, more advanced nuclear technologies. I greatly appreciate the testimony offered today by Dr. Mark Peters of the Idaho National Lab which, as DOE's lead Nuclear Energy Laboratory, is doing phenomenal work in the area of nuclear energy technologies.

I would also highlight the Oak Ridge National Laboratory (ORNL), home to Alvin Weinberg and Admiral Hyman G. Rickover, both pioneers in the development and use of nuclear energy. ORNL has established itself as an influential leader in the advancement of nuclear technology. Southern is proud to be partnering with ORNL for the DOE-awarded research project involving molten chloride fast reactor technology and we commend ORNL's role in supporting the use of nuclear technology for the nation's security as well as commercial interests. ORNL's efforts over many decades have resulted in development and operation of 13 nuclear reactors. Using nuclear energy for a host of applications, from fueling commercial nuclear power plants to powering nuclear submarines, ORNL has demonstrated the power of nuclear energy to protect national security and to drive economic growth. We salute ORNL's achievements and its commitment to build on Alvin Weinberg's "notion of a laboratory whose mission evolves and strengthens over time."[6]

In fact, the vision for the Molten Salt Reactor (MSR) technology was originally developed at ORNL. This technology, regrettably dismissed many decades ago for political and other reasons, is now benefitting from renewed national and international interest. MSR technology would have passive safety features to ensure safe operation without human or mechanical intervention, a considerably smaller nuclear waste profile, more efficient fuel use, and lower construction costs. As a range of technology options are explored, we will advocate for and encourage industry-led collaboration with DOE, vendors, utilities, universities and national labs to leverage capabilities and share some of the risks.

---

[5] Third Way, *The Advanced Nuclear Industry* (June 15, 2015), available at http://www.thirdway.org/report/the-advanced-nuclear-industry.

[6] Oak Ridge National Laboratory, *History*, available at https://web.ornl.gov/ornlhome/history.shtml.

Modernizing the Licensing Framework for Advanced Reactors

A common element of all discussions about advanced reactors is the need to modernize the licensing framework to accommodate different kinds of reactors.[7] Our current regulatory framework for the licensing of nuclear power plants has its roots in the federal government's initial efforts to promote commercial nuclear power after the passage of the Atomic Energy Act of 1954 (the "AEA") when the Atomic Energy Commission ("AEC") began to encourage the development of commercial nuclear power production in the private market. The federal government helped spur innovation and investment in nuclear power production through research and development efforts such as test reactors and laboratories that would eventually share information with the private nuclear power industry. At the same time, the federal government provided economic assistance to those private companies willing to take the first steps to construct and license nuclear power plants. The AEC and the private sector researched and experimented with several different types of reactors, including light-water reactors, salt-cooled reactors, and fast-breeder reactors.

Prompted by the backing of the AEC, the commercial nuclear power industry started to take shape, and the United States led the world in innovation as the nuclear industry grew rapidly throughout the 1960s. Eventually, the AEC and industry focused on light-water reactor technology, and the federal government's reactor licensing framework grew up around, and was molded to fit the needs of, light-water reactor designs, resulting in a workable licensing process in which the nuclear power industry could remain generally assured of the regulatory framework for its investment.

With the passage of the Energy Reorganization Act of 1974, the AEC was abandoned, and its dual functions of regulating the nuclear power industry while simultaneously promoting nuclear power were split among the Nuclear Regulatory Commission (NRC) and the Energy Research and Development Administration (ERDA), respectively. In 1977, ERDA's functions were transferred to DOE, an agency deserving of credit for much of the innovation in commercial nuclear power after the passage of the Act. Most of DOE's nuclear facilities and programs are exempt from NRC regulation, allowing DOE to research and develop technologies that may otherwise remain unexplored. Consequently, much of the research and development in the nuclear power industry hinges on decisions of the federal government.

---

[7] In a recent article, MIT professor Richard Lester articulates a vision for a "new roadmap for nuclear innovation in the United States," forecasting "three successive waves of advances" including: (1) a first wave breaking during the next decade that will enable extended operating lives for the nation's existing fleet of nuclear reactors; (2) a second wave occurring in the 2030-2040 timeframe, which he describes as a "critical period" for "rapid scale-up of nuclear energy"; and (3) a third wave occurring in the post-2050 period with even further advances. Lester, Richard K. "A Roadmap for U.S. Nuclear Energy Innovation." *Issues in Science and Technology* 32, no. 2 (Winter 2016). The kinds of actions and ideas that Dr. Lester suggests align fairly well with those that I and others in the industry believe are needed to facilitate nuclear innovation in America. In particular, his roadmap calls for changes within the NRC, an expanded role for the national labs, and further support of international collaboration at DOE and elsewhere. Dr. Lester also correctly cautions that it is "premature at this stage to attempt to identify a winner among all [types of nuclear] innovations…" I would commend Dr. Lester's article to this Committee for review and consideration.

As the NRC adopted the AEC's regulatory duties, it continued to implement the reactor licensing process adopted by the AEC that required a licensee to first obtain a construction permit and then an operating license at a later point in time. Throughout the 1970s and 1980s, utilities constructing nuclear plants struggled to complete projects on time and experienced cost overruns associated with evolving licensing requirements in the two-step process. These licensing inefficiencies were magnified by new requirements imposed in response to the Three-Mile Island accident in 1979. Eventually, the financial and regulatory risks associated with development of commercial nuclear power plants resulted in an effective moratorium on construction. Technical innovation across the nuclear power industry slowed markedly and the infrastructure for the manufacture of materials and components needed for the construction of new plants in the United States was severely diminished.

In an effort to mitigate these difficulties, the NRC developed a new, combined construction and operating licensing process, codified at 10 C.F.R. Part 52, which allowed for the resolution of design and environmental licensing requirements prior to the start of construction. The Part 52 process provides for as much regulatory oversight and ensures the safety of the public every bit as much as the old two-step approach but with more regulatory stability and predictability, which encourages investment in commercial plants. Once again, the actions of the federal government spurred new research and investment in nuclear reactor technologies. The Part 52 licensing process paved the way for the construction of new nuclear reactors utilizing new reactor technologies, such as the Westinghouse AP1000 reactor that will be in operation at Plant Vogtle Units 3 and 4. The Energy Policy Act of 2005 further stimulated investment in new nuclear reactor technology through federal incentives such as tax credits and loan guarantees. Because of regulatory improvements implemented by the NRC and other means of support from the federal government, the nuclear power industry in the United States stood on the brink of what many called the "nuclear renaissance."

Today, the nuclear power industry stands at yet another crossroads. Commercial nuclear power is expanding across the world yet the United States is not currently at the center of the technological innovation driving much of the expansion. While the Part 52 licensing process proved beneficial to the industry, the fact that it, like the initial two-step licensing process, is based on light-water reactor technology limits its efficacy for the licensing of Gen-IV reactors. The current regulatory framework with its inefficient, exemption-based licensing approach will be ineffective for licensing non-light water reactor technology. Current procedures would require potential investors to spend billions of dollars without a defined path for licensing a Gen-IV reactor. The NRC, in a 2012 report to Congress, outlined the need to develop a regulatory approach that "supports the unique aspects of advanced designs" and includes, among other things, "identifying policy, technical, and licensing issues" and "developing the regulatory strategies to support efficient and timely reviews" for advanced reactors.[8] The NRC specifically identified the need to streamline its application process by developing a "new, risk-informed, performance-based regulatory structure for non-LWR advanced reactor designs."[9] We agree.

---

[8] *See* NRC, Report Congress: Advanced Reactor Licensing, at p. iv (August 2012), available at http://pbadupws.nrc.gov/docs/ML1215/ML12153A014.pdf.

[9] *Id.*

Likewise, the NRC recognized that an advanced reactor design that "uses fuel that differs significantly from the current [fuel] type (zirconium-clad, low-enriched uranium dioxide) will require the evaluation of technical and regulatory approaches to the licensing of fuel fabrication, transportation, storage, and waste disposal operations."[10] A modernized regulatory framework that effectively addresses the needs associated with licensing a non-light-water reactor will signal to the private sector that it can invest in research and development of advanced reactors knowing that the licensing environment does not favor a single technology, thereby allowing various kinds of technologies to be developed and licensed.

When developing a licensing framework that can work for advanced reactors, I would endorse the "triple A" approach. That is, where existing regulations are appropriate, "adopt" them; where simple changes are needed to modify existing rules in order to make them a better fit for advanced reactors, "adapt" them; and where the characteristics of advanced reactors require new regulatory structures and programs, "advance" them. In all respects, the NRC – as the safety regulator – should determine the required safety performance metrics, while the industry and its partners should focus, through consensus standards organizations, on developing the "how" to comply with performance standards and design requirements. By doing so, we can prevent stagnation in the development of advanced reactor designs and ensure that the newest, safest, and most efficient nuclear reactors will be built in the United States.

While the licensing framework is improved, Congress will be called upon to provide adequate funding for the NRC to fulfill its responsibility in this regard. As this Committee is aware, ninety percent of the NRC's budget is currently derived from industry fees charged primarily against the nation's fleet of existing reactors. Cost associated with funding advanced reactor licensing improvements and related activities should not be borne by existing reactors, and I would encourage Congress to consider robust funding of advanced reactor programs and activities.

Advanced Reactor Working Group (ARWG)

At the outset of my testimony, I mentioned NEI's Advanced Reactor Working Group, which has representatives from seven electric utilities and ten reactor design companies. Over the next 30 years, a significant amount of the existing generating capacity will be retired. The ARWG was created with the understanding that decisions as to what technologies will replace recent and upcoming nuclear reactor retirements will be made within the next 10-20 years. In the short- to medium-term, light water reactors will remain the dominant and most economic means of electricity production from nuclear energy, but decisions about future energy investments will most certainly take into account the contributions of advanced non-light water reactors.

With this reality in mind, the ARWG is charged with developing an industry vision of a sustainable program to support the development and commercialization of advanced reactors, ultimately with commercial availability in the 2035-2040 timeframe. Currently, the working group is focused on:

---

[10] *Id.* at v.

(1) Developing, communicating, and implementing an industry strategic plan for the development and commercialization of advanced reactor technologies.

(2) Developing legislative proposals at the federal level to appropriately support development of advanced reactors.

(3) Identifying and proposing changes to the NRC licensing framework for advanced reactors.

(4) Establishing a demonstration program for construction and operation of multiple advanced reactor designs at a DOE site, or utility site, or a yet to be developed test center.

ARWG looks forward to serving as a resource to this Committee, DOE, and other stakeholders.

Recent Positive Steps by the Federal Government

The ARWG is not the only new entity focused on advanced reactors. We also applaud the Secretary of Energy's Advisory Board (SEAB) Task Force on the Future of Nuclear Power, which has recognized that nuclear power is an "important carbon-free power source for the U.S. and the world." The SEAB Task Force, which has been proactively engaging with the nuclear industry, sees the need to explore the kinds of nuclear reactors that should be deployed in the 2030-2050 timeframe. As an industry, we look forward to the Task Force's final report at the end of this year and reviewing their findings and recommendations for achieving technical milestones on advanced reactor designs, certification, engineering, prototype testing, licensing, and deployment.

Similarly, in November of last year, the Administration announced a new program called "Gateway for Accelerated Innovation in Nuclear" (GAIN), which is intended to "provide the nuclear energy community with access to the technical, regulatory, and financial support necessary to move new or advanced nuclear reactor designs toward commercialization while ensuring the continued safe, reliable, and economic operation of the existing nuclear fleet." A key element of the GAIN initiative is to provide all nuclear stakeholders with a "single point of access" to the array of federal assets and programs including the DOE complex and national labs. Whether Congress, DOE, the states, or at the international level, a consensus exists that nuclear power should have a central role in meeting the world's energy demands into the future.

In addition, the Blue Ribbon Commission (BRC) noted in its final report that the benefits of advances in nuclear energy technology justifies sustained public and private-sector support for research and development into advanced reactor and fuel cycle technologies. Furthermore, the BRC strongly recommended increased effort in developing a regulatory framework for advanced nuclear technologies to help guide design research and lower barriers to commercial investment by increasing industry confidence that advanced reactors can be effectively licensed.

Finally, I applaud Congress for its continued interest and support of advanced reactor initiatives. Members of Congress on both sides of the aisle have been working proactively to promote the next generation of nuclear power and substantial legislative progress has been made particularly in the last few months. For example, the House of Representatives recently passed the Nuclear Energy Innovation Capabilities Act (H.R. 4084), a bipartisan bill to support federal research and development and stimulate private investment in advanced nuclear reactor technologies. A similar bill, S. 2461, introduced by Senator Crapo, was approved by the Senate in January with overwhelming support as an amendment to the bipartisan energy bill (which was ultimately passed by the Senate on April 20, 2016). Most recently, Senator Inhofe, Chairman of the Senate Environment and Public Works Committee, introduced bipartisan legislation with Senators Booker, Crapo, and Whitehouse to support advanced reactor licensing and to reform the NRC's budget and fee structure. We are deeply grateful for the investment of time and resources that this Committee and others in Congress are putting into this important national priority.

I would also highlight the important decision that Congress made just last year to fund the DOE Advanced Reactor Concepts (ARC) program for FY2016, including funds for "continued development of two performance-based advanced reactor concepts."[11] ARC supports the "research of advanced reactor subsystems and addresses long-term technical barriers for the development of advanced nuclear fission energy systems utilizing coolants such as liquid metal, fluoride salt, or gas." We are pleased that the Senate and House Fiscal Year 2017 Energy and Water Appropriations bills both include funding increases for the ARC program. We recognize Senators Alexander and Feinstein for their leadership in making advanced nuclear technologies a funding priority in the FY2017 Energy and Water Development Appropriations bill.

Conclusion

Other nations are investing in nuclear technologies, and a key question is whether the United States will be the global leader in nuclear energy in the future as we have been in the past. Today, over 40% of the nuclear power plants under construction globally are located in China. With a significant civilian nuclear capability, Russia is also taking a close look at advanced reactors, and France and other European Union members are currently working on several other advanced reactor designs. Our nation should not cede nuclear innovation to others.

Decades ago, senators looked to Admiral Rickover for vision and expertise on the potential for civilian nuclear power. He once said: "We must live for the future of the human race, and not for our own comfort or success. We realize that continuous change is difficult; it takes us out of our comfort zone. But our future existence relies on change, and the most impactful change comes through innovation." With the current fleet of LWRs and operation of previous advanced test reactors, it took just a decade to progress from concept to commercial operation. This required innovation and collaboration. As we look ahead to Gen-IV reactors, I believe our nation has the knowledge and expertise to make this kind of technological progress in nuclear energy happen again. As before, innovation and collaboration will provide the keys to success.

Thank you for allowing me to appear before this Committee today. I will be glad to answer any questions you might have.

---

[11] *See* Explanatory Statement, Division D, at p. 29.

The CHAIRMAN. Thank you, Mr. Kuczynski.
Dr. Peters, welcome to the Committee.

STATEMENT OF DR. MARK PETERS, DIRECTOR, IDAHO NATIONAL LABORATORY

Dr. PETERS. Thank you. Thank you, Chairwoman Murkowski.

Thank you, Chairwoman Murkowski and Ranking Member Cantwell and members of the Committee for this opportunity to speak to you today. I am Mark Peters, Director of the Department of Energy's Idaho National Laboratory (INL), the lead national laboratory for nuclear energy.

I'm pleased to participate in this most distinguished panel before the Committee, and I'll request that my written testimony be made part of the record.

The CHAIRMAN. It will be included.

Dr. PETERS. Before I begin my testimony, I'd like to thank Senator Risch for his support of INL and for co-sponsoring the Nuclear Energy Innovations Capability Act, authored by Idaho Senator Crapo, Senator Booker and Senator Whitehouse. This legislation, as part of the Senate Energy bill, is the companion to the House measure of the same name. The House and Senate legislation are important enablers to much of what I will discuss today.

The U.S. is widely recognized as a world leader in the development of advanced nuclear reactors; however, leadership is earned, not granted and other nations are investing to develop the facilities, capabilities and people necessary to excel. The U.S. has the opportunity to regain domestic manufacturing and supply chain capabilities lost when we did not build new reactors during the last 30 years.

Small modular reactors and advanced nuclear reactors can be entirely sourced in the U.S. creating new advanced manufacturing facilities vital for economic growth.

The value proposition for U.S. nuclear energy has never been stronger. There are strong global and domestic interest in nuclear energy due to the recognition that safe, secure, reliable and affordable energy is the engine for economic growth, prosperity and quality of life.

The U.S. cannot meet increasing electricity demand and stringent clean air goals with renewable energy alone. In this effort nuclear and renewables become complementary. Safe and reliable nuclear energy provides 19 percent of total electricity and 63 percent of the U.S. electricity sector's carbon-free generation today.

We are on the cusp of a fundamental transportation in nuclear energy. The existing light water reactor fleet will serve as a bridge to SMRs and advanced reactor technologies. We have developed tremendous expertise in operating light water reactors at the highest level of safety and efficiency and that expertise is relevant to advanced reactor design and operations.

Our conversations with advanced reactor developers indicated challenges in two main areas: resolving technical and licensing challenges at an early stage and addressing remaining technical licensing and economic questions at the demonstration and deployment phase.

In November the Administration announced formation of the Gateway for Accelerated Innovation in Nuclear, the GAIN Initiative. GAIN will provide the people and facilities to develop and test key components and mitigate uncertainty while allowing developers to fine tune their designs.

Resolving technological risks can give investors the confidence to move forward with increased financing for an advanced reactor design. As a core part of GAIN, INL and our national laboratory and university partners are the portal for designers and developers interested in a wide range of DOE nuclear energy related capabilities and expertise.

For the later stage commercialization, companies may access INL and other government sites for demonstration and deployment capabilities which can reduce costs and improve performance of their design as it moves to full commercialization.

An example of this is the recent announcements related to potential siting of a NuScale SMR in the INL site which will provide a near-term opportunity to provide a demonstration platform for innovative nuclear technologies.

New types of industry partnerships are also driving strategic investments. For instance, INL is working cooperatively with TerraPower to build new capabilities to provide R and D support for the traveling wave reactor concept Dr. Gilleland discussed earlier.

The experimental fuels facility at INL's materials and fuels complex has gone a transformation in recent years to expand capabilities for the traveling wave reactor.

Madam Chair, we are creating a new paradigm in nuclear innovation and nuclear energy. This paradigm involves new ways of working with the diverse nuclear community that includes utilities, startups, large nuclear suppliers, government entities, non-profits and everyone on this Committee. We are your national laboratories, and we're open for business.

Recent White House and DOE initiatives as well as congressional legislation and funding are setting the stage for research, development and deployment of new and advanced nuclear energy systems. The end result will mean cleaner, more plentiful energy with the potential for lifting billions out of energy poverty.

Thank you for inviting me to testify, and I look forward to your questions.

[The prepared statement of Dr. Peters follows:]

Name: Dr. Mark Peters
Title: Director
Organization: Idaho National Laboratory

Nuclear Energy Innovation - The Path Forward for Small Modular Reactors and Advanced Nuclear Reactors

Introduction

Thank you Chairwoman Murkowski and Ranking Member Cantwell, and members of the Committee, for this opportunity to speak to you today. I am Dr. Mark Peters, Director of the DOE Idaho National Laboratory (INL) since October 1. I most recently served as Associate Laboratory Director at Argonne National Laboratory (ANL) in Illinois. I am pleased to participate in this most distinguished panel before the Committee. I request that my written testimony be made part of the record.

Before I begin my testimony, I would like to thank Idaho Senator Risch for his continuing support of INL and the research conducted at the national laboratories and for becoming a cosponsor of the Nuclear Energy Innovation Capabilities (NEICA) Act, S. 2461, authored by Idaho Senator Crapo, Senator Booker (D-New Jersey), and Senator Whitehouse (D-Rhode Island), which, on an 87-4 vote, was included as a bi-partisan amendment to the Senate Energy Bill last month.

The legislation is the companion to the House measure of the same name, H.R. 4084, offered by Representatives Lamar Smith (R-Texas), Eddie Bernice Johnson (D-Texas), and Randy Weber (R-Texas). It passed the House Science Committee and was recently added to H.R. 4909, the National Defense Authorization Act. The House and Senate legislation are important enablers to much of what I will discuss today.

Laying the Groundwork for a New Nuclear Innovation Paradigm

INL has a vision for the vital role nuclear energy must play as part of the future global energy system. The U.S. is widely recognized as a world leader in the development of advanced nuclear reactors, as demonstrated by the numerous cooperative research agreements and partnerships in place between DOE and foreign partners. However, leadership is earned, not granted, and other nations are investing to develop the facilities, capabilities, and people necessary to excel.

Light-water small modular reactors (SMR) and advanced reactors also provide the opportunity to re-establish the domestic nuclear industry (entire value chain) – a key to global leadership. The U.S. has the opportunity to regain domestic manufacturing and supply chain capabilities lost when we did not build new reactors during the last 30 years. SMRs and advanced nuclear reactors can be entirely sourced in the U.S. creating new advanced manufacturing facilities vital for economic growth. The proposed NuScale SMR, which could be built on the INL Site, would create thousands of jobs during the construction phase and hundreds of permanent jobs with annual incomes far above the regional, state, and national averages. According to the Idaho Department of Labor, this project would infuse millions of dollars annually into the local and state economies. This is but one example of why the U.S. cannot afford to fall behind and lose a competitive advantage to other nations.

The value proposition for U.S. nuclear energy has never been stronger. Recognizing safe, secure, reliable, and affordable energy as the engine for economic growth, prosperity, and quality of life, there is strong global and domestic interest in nuclear energy. Concerns about climate change and the associated need to limit the GHG emissions, and reliability of clean energy supply, also drive the increasing interest in this technology. During the recent international negotiations on greenhouse gas emissions, the White House declared nuclear energy a key component of U.S. strategy to achieve carbon reduction goals. The U.S. cannot safely, securely, and affordably meet increasing electricity demand and stringent clean air goals with renewable energy alone. In this effort, nuclear energy and renewables become complementary. Safe and reliable nuclear energy provides 19 percent of total electricity and 63 percent of U.S. electricity sector's carbon-free generation today. Nuclear energy's contribution to clean air must be maintained and even grow into the future.

We have immediate challenges and opportunities before us to enable SMRs and advanced reactors. We set about determining what could be done last March. With our partner national laboratories and the nation's universities, we hosted a set of six simultaneous workshops - the Nuclear Innovation Workshops - aimed at developing creative solutions to accelerate innovation in nuclear energy. We asked the private sector: what are your needs?

Our conversations with advanced reactor developers indicated challenges in two main areas: resolving technical and licensing challenges at an early stage and addressing remaining technical, licensing, and economic questions at the demonstration and deployment phase. In fact, one of the top recommendations to come out of the workshops was the need for a national test bed – or test beds – where the developers working on advanced nuclear technologies can mature the design of their system and associated components.

Then, in November, the Administration announced formation of the Gateway for Accelerated Innovation in Nuclear (GAIN). As the nation's lead nuclear energy laboratory with distinctive capabilities, expertise, and facilities, INL, in collaboration with ANL and Oak Ridge National Laboratory (ORNL), will play a leadership and integration role in the multi-lab and university implementation of GAIN.

GAIN, as a research, development, and demonstration platform through public-private partnership, will provide the people and facilities to develop and test key components and mitigate uncertainty while allowing developers to fine-tune their design. Resolving technological risks can give investors the confidence to move forward with increased financing for an advanced reactor design. Building on the success of national user facilities across its nationwide national laboratory complex, DOE has made great strides in making its unique facilities, capabilities, and people available to university and industry research partners, and GAIN will take this effort to a new level.

As a core part of GAIN, INL and our partners is the portal for designers and developers interested in a wide range of DOE nuclear-energy related capabilities and expertise. We are developing a detailed execution plan for GAIN with input from investors and industry. The execution plan will include streamlined contracting processes with industry, a communication plan, and describe the organizing principle for the DOE Office of Nuclear Energy (DOE-NE) sponsored relevant R&D programs. We have already made considerable progress in putting a number of critical elements of GAIN in place. We have established a National GAIN organization, including an executive advisory committee comprised of leaders from industry and the national laboratories. Technology specific workshops with various stakeholders are planned in the near future with the objective of streamlining the base DOE R&D programs informed by industry and investor needs. We have developed and proposed to DOE-NE, a five-year DOE-NE funding plan consistent with GAIN strategic objectives.

In late January, the Third Way with INL and partner national laboratories hosted an advanced nuclear energy summit to engage the private sector in regard to creation of new opportunities for collaborative research and development in light-water SMRs and advanced, non-light water reactor technologies. I would like to thank Chairwoman Murkowski for her remarks, which were heard by the hundreds of people attending the event. Advanced reactors may be suitable for micro grids like those found in Alaska. In many communities around the world, where there are impediments to building large-grid infrastructure, there continues to be great enthusiasm for the development of innovative nuclear energy technologies to achieve the full benefits of this clean energy source.

Light water reactors have achieved unparalleled safety and environmental milestones. However, in the light of growing demand for clean energy nationally and globally, there is an urgent need to develop and deploy significant new and flexible nuclear energy capacity, starting now and ramping up significantly in the 2030-2040 timeframe. The next generation of reactors will provide enhanced passive safety features, while increasing efficiency and reducing environmental impact. They will also prove to be more economical and reduce the proliferation risk. This requires accelerated innovation, and commercialization of SMRs and advanced reactors.

Gateway for Accelerated Innovation in Nuclear (GAIN)

The Administration's GAIN proposal has four components: 1) access to capabilities (test facilities, high performance computing, national experts, and demonstration capabilities); 2) a nuclear energy infrastructure database; 3) small business vouchers; and, 4) assistance in navigation of the regulatory process.

INL, with partner universities and other national laboratories, will facilitate an R&D test bed for industry to access laboratory expertise and nuclear facilities and capabilities. Resolving technological risks can provide confidence to investors to move forward with increased financing for a reactor design. For the later-stage commercialization, companies may access INL and other government sites for demonstration and deployment capabilities, which can reduce costs and improve performance of their design as it moves to full commercialization. An early example of this is the recent announcements related to potential siting of a NuScale SMR on the INL Site. The new cost of future reactor technologies has been notoriously difficult to predict. One function of the GAIN initiative is to remove engineering uncertainty for the construction process and allow developers to provide a better estimate of costs. Through GAIN, we will address different models for partnerships and incorporate what has been proven to work well into our new paradigm for advanced reactor development and deployment.

DOE continues to invest in maintaining and developing state-of-the art research capabilities to more effectively and efficiently develop innovative nuclear energy technologies. Recent examples include improved world-leading post-irradiation examination capabilities, the restart of transient testing capabilities, and high-fidelity modeling and simulation capabilities. These, and other capabilities, will be made available through GAIN for retirement of technical, licensing, and financial risk associated with the critical components of innovative designs.

DOE continues to invest in the experimental and computation capabilities in support of nuclear technologies. This includes the added capabilities at the Office of Science User Facilities, such as Spallation Neutron Source (SNS), National Synchrotron Light Source (NSLS-II), Advanced Photon Source (APS), and Linac Coherent Light Source (LCLS) to conduct experiments with nuclear materials. One major missing capability in the complex is a fast-spectrum test reactor needed for accelerated irradiation effect studies for fission and fusion systems, and the development and qualification of fuels and materials to be used in future advanced reactors. DOE is evaluating the needs for such a facility. Addressing the needs for a new fast-spectrum irradiation testing capability is also included in the recent House and Senate energy authorization bills, H.R. 4084 and S. 2461, respectively.

DOE is publishing the Nuclear Energy Infrastructure Database (NEID), which provides a catalogue of existing nuclear energy related experimental and computing infrastructure that will enhance transparency and support nuclear community engagement through GAIN. According to DOE, "NEID currently includes information on 802 research and development instruments in 377 facilities at 84 institutions in the United States and abroad. Nuclear technology developers can access the database to identify resources available to support development and implementation of their technology, as well as contacts, availability, and the process for accessing the capability."

DOE also announced a small business voucher (SBV) program to support the strong interest in nuclear energy from a significant number of new companies who are working to develop advanced nuclear energy technologies. According to DOE, the government plans to make funding "available in the form of vouchers to provide assistance to small business applicants (including entrepreneur-led start-ups) seeking to access the knowledge and capabilities available across the DOE complex." This program is modeled after the successful SBV program at the DOE Office of Energy Efficiency and Renewable Energy, but tailored to the unique needs of the nuclear industry. In 2016, a small pilot program is being executed for a total of $2 million in funding. The technical reviews for the many applications have been completed, and DOE soon will be making the final decisions for the awards.

DOE will work through GAIN with prospective applicants for advanced nuclear technology to better understand and navigate the NRC regulatory process for licensing new reactor technology. At the September 2015 workshop on advanced non-light water reactors, the NRC said it would: 1) consider whether a "staged" licensing process is possible; 2) clarify some targeted guidance of particular interest to designers, such as prototype guidance; and 3) continue progress on identified policy issues. A staged licensing process, in particular, would inform further investments by the private sector as technical and licensing risks are retired. The second DOE/NRC workshop is scheduled next month to address these issues further.

We are creating a new paradigm in nuclear innovation and nuclear energy. This paradigm involves new ways of working with a diverse nuclear community that includes utilities, startups, large nuclear suppliers, government entities, non-profits, and everyone on this Committee. We are your national laboratories, and we are open for business.

DOE's Lead Laboratory for Nuclear Energy: Idaho National Laboratory

For more than 60 years, INL has played an important leadership role in the development and deployment of nuclear energy and more recently the development of next generation nuclear reactors. It is an exciting time for nuclear energy as the world looks at this essential component to address energy security and the risks presented by a changing climate. Nuclear energy plays a vital role as part of the future global energy system. The value proposition for nuclear energy has never been stronger.

INL is an applied research and development laboratory. We work to enable innovation to ensure that secure and reliable advanced nuclear energy technologies are available to the U.S. and a global energy market. The laboratory is working with federal agencies (including the NRC), universities, and other national laboratories to establish and maintain a domestic nuclear energy capability. This work is anticipated to culminate in the U.S. providing global leadership and reestablishing a supply chain for advanced nuclear energy systems development and deployment.

INL is in a prime position to take on national challenges and seize opportunities in supporting nuclear energy. The work INL conducts in safety, nonproliferation, and the economics of nuclear energy supports the global nuclear energy market. INL researchers work to develop technology solutions to address climate change, ensure secure and resilient energy infrastructure, enable nuclear material security, and sustain U.S. leadership in a competitive environment.

INL has formed advanced reactor partnerships with the private sector. Currently the laboratory is working on: 1) advanced reactor design evaluation; 2) hybrid nuclear energy systems development; 3) digital instrumentation and control work; 4) innovative fuels and materials design development, fabrication and demonstration; and 5) modern and creative risk analysis techniques.

One of the most important projects undertaken recently by INL, and authorized by the 2005 Energy Policy Act, was the Next Generation Nuclear Plant (NGNP). This high-temperature gas reactor was under development as part of the GENIV nuclear energy program. From 2005-2013, the NGNP program worked to address and resolve critical technical and licensing issues. In 2014, the NRC staff released an assessment report that found the NGNP technical approach on fuel qualification was reasonable. The laboratory is still working with the NRC to qualify the fuel.

However, the (NGNP) experience also showed us some of the challenges associated with licensing advanced reactor technologies. We strongly believe INL's engagement with the NRC and industry on a non-light water reactor technology licensing effort will be a valuable part of enabling commercialization of advanced reactors.

The Transient Reactor Test Facility (TREAT) was specifically built at the INL site to conduct transient reactor tests where the test material is subjected to neutron pulses that can simulate conditions ranging from mild upsets to severe reactor accidents. The reactor was constructed to test fast reactor fuels, but has also been used for light-water reactor fuels testing as well as other special purpose fuels (i.e. space reactors). Renewed interest in this facility was sparked by the nuclear accident in Japan in 2011 when Congress directed DOE to renew efforts to develop more accident-tolerant nuclear fuel. We are on track to restart this important test facility by 2018.

In the last 10 years, we have made considerable progress in upgrading the capabilities at the Materials and Fuels (MFC) complex. Now, the facility includes state-of-the-art capabilities for laboratory-scale fuel fabrication, fuels and materials characterization, post-irradiation examination and analytical chemistry. The research output continues to improve. Likewise, we are making considerable progress in managing the Advanced Test Reactor (ATR) reliability and preventive maintenance issues, and increasing the efficiency of this world-class test reactor. We look forward to operating the ATR until 2050 and beyond. The reactor is operating on a much more predictable schedule, compared to previous years. We are also investing considerable R&D dollars for in-pile instrumentation for state-of-the-art measurements during irradiation experiments.

INL also is leading the way in a promising application of advanced reactors as part of hybrid energy systems that use heat from fission to power industrial processes that would otherwise rely on fossil fuel/natural gas to provide heat. Another potential use is hydrogen production. Large quantities of feedstock are being used in the petrochemical industry to upgrade heavy crude oil and produce fertilizers. We are working cooperatively with the National Renewable Energy Laboratory (NREL) to examine new ways to integrate renewable energy with nuclear energy systems. In addition, we have established a 91,000-square-foot Energy Systems Laboratory (ESL), located on the INL's Idaho Falls Research and

Education Campus, for research and development in bioenergy, hybrid energy systems, and advanced vehicle and battery testing programs.

INL also leads a robust DOE program in collaboration with industry in support of Light Water Reactor (LWR) technology - the Light-Water Reactor Sustainability (LWRS) program. For example, the laboratory has partnered with Arizona Public Service's Palo Verde Nuclear Generating Station to design a modernized control room for operating commercial reactors. This work supports the long-term sustainability and efficiency of the nation's existing nuclear reactor fleet by assisting nuclear utilities to address reliability and obsolescence issues of legacy analog control rooms.

Finally, INL and partner national laboratories support the ongoing and important efforts by DOE and the nuclear industry to develop and commercialize SMRs. We are providing direct technical support to companies such as NuScale as part of their design and licensing process and also are looking to provide technical support and infrastructure to deploy SMRs in the United States. We are a proud partner with NuScale and are working together with DOE and the Utah Associated Municipal Power Systems (UAMPS) to build the nation's first SMR at the INL.

Nuclear Energy Innovation is Strengthened at Idaho National Laboratory

The new nuclear energy innovation paradigm is driving new types of partnerships between national laboratories, universities, and industry. The laboratory is becoming an entry point, or portal, for utilities, large nuclear suppliers, entrepreneurs, and small businesses to access the national laboratory facilities and staff expertise. The idea for user facilities originated with the concept that exceptional research capabilities and expertise developed to meet national challenges should also be made available to non-government users. Idaho National Laboratory's Nuclear Science User Facilities (NSUF) is one of a number of DOE user facilities in the United States and the nation's only designated nuclear energy user facility. As the only DOE nuclear energy user facility with partner facilities, the NSUF is the hub that connects a broad range of exceptional nuclear research capabilities spanning the U.S.

The nuclear innovation test bed concept is a promising development in advanced reactor research and development. In a recent example, INL is working cooperatively with TerraPower to build new capabilities and provide support for the Traveling Wave Reactor (TWR) concept, a new type of fast reactor, under multiple cooperative research and development agreements (CRADA). In fact, the Experimental Fuels Facility (EFF) at INL's Materials and Fuels Complex has undergone a transformation in recent years to expand capabilities for the TWR. In addition to the extrusion press, the EFF has been newly equipped with a billet-casting furnace, inert glove box, draw bench, annealing furnace, and several other types of machining equipment.

INL is also partnering with other national laboratories to expand the capabilities of pre-existing DOE research programs. In December, DOE's Consortium for Advanced Simulation of Light Water Reactors (CASL), led by Oak Ridge National Laboratory (ORNL), agreed to help NuScale establish new cost-shared modeling and simulation tools for its SMR design. Through this agreement, CASL tools will be expanded to better simulate SMR operations and inform design decisions. This research will allow for a more efficient reactor design that will enable better operation lifetime.

In another recently formed public-private partnership, on January 15, DOE announced INL would partner with X-energy to address fuel development challenges of the Xe-100 Pebble Bed Advanced Reactor. Other partners include: BWX Technology, Oregon State University, Teledyne-Brown Engineering, SGL Group, and ORNL. This award provides an example of the public-private partnerships envisioned under the recently launched GAIN initiative. DOE's investment will be $6 million this year. The possible multi-year cost-share value for this research is as much as $80 million.

Summary

The outstanding safety record of the U.S. nuclear power industry is a direct result of the groundbreaking research, development, and demonstration of the technologies and safety systems developed in the last century. This leadership position enabled the U.S. to set the terms and act as a key resource for other nations as they developed reactors for power and research, and to ensure the security and eventual return of fuels and materials that could otherwise have been diverted.

We are on the cusp of another fundamental transformation in nuclear energy. The existing light water reactor fleet will serve as a bridge to SMRs and advanced reactor technologies. We have developed tremendous expertise in operating LWRs at the highest levels of safety and efficiency and much of that expertise will be relevant to advanced reactor design and operations.

We will see today's light-water reactors, SMRs, and advanced non-light-water reactors operating side-by-side. When looking at siting nuclear reactors, it may make sense to prioritize existing sites that have the infrastructure and workforce in place to support a new reactor and, moreover, a good deal of site characterization has already been performed. One key role that INL and the national laboratories can play is to provide the technical foundation for those siting decisions and platforms for demonstration (as part of GAIN).

I am very pleased to be at the helm of INL, the lead national laboratory for nuclear energy, with more than 4,000 employees devoted to science and innovation in advanced nuclear energy technologies. Recent White House and DOE initiatives, as well as Congressional legislation, are setting the stage for research, development, and deployment of new and advanced nuclear energy systems. The end result will mean cleaner, more plentiful energy with the potential for lifting billions out of energy poverty.

Thank you for inviting me today to testify, and I look forward to your questions.

The CHAIRMAN. Thank you all, I appreciate the testimony this morning and your leadership in these areas.

Thinking about the exciting developments and the prospects, and Dr. DeWitte, your focus on micro-reactors is really quite exciting for a place like Alaska where we are not connected to anybody else's grid. You have to have some level of scalability. You have to be able to take it down to small communities or perhaps with our military installations. I look at this and see there is room for extraordinary potential.

As with any good idea though, you have situations relating to the financing and you have a regulatory process; so I want to start with my questions this morning on the regulatory aspects.

Mr. Hopkins, you probably have as much experience as anybody with NuScale in terms of the process that you have gone through. You have been underway for about 15 years working with the NRC since 2008. As you have outlined it, my take away is you think that you are able to work within the NRC licensing process as it is.

Mr. Kuczynski, you have suggested that there are some real challenges within the process as it exists today taking these new technologies and basically trying to make them fit within a structure that has been designed for basically a different model or a different approach.

I would like to have a more full discussion on this issue of whether or not we need to see this licensing framework restructured, to be more adaptable to recognize that we have a whole range of different technologies that we are considering right now. Recognizing that time is money, what do we do with the regulatory process as it exists now?

I kick that out for general discussion.

Mr. Hopkins, Mr. Kuczynski, go ahead.

Mr. HOPKINS. Senator, we have been working with the NRC, since 2008. However, there are differences between our small modular reactor and a large plant. There are exceptions.

As an example, we don't have any hydrogen production because we're oxygen starved in our reactors; therefore, we don't need a hydrogen re-combiner.

So there are things that we are working with the NRC, which takes a lot of time and effort, to get them understanding what those differences are. We typically do what's called topical reports. So any of these variances that we see we write a technical paper for review by the NRC. And so far, I think we've submitted eight or nine of these topical reports. I believe we have 15 to submit.

But we view our opportunity here, as was mentioned earlier, to help pave the way for the next Gen IV advanced reactor because as the NRC stays focused and understands the nuances and change. Although we are a light-water reactor, which they've known the technology for 50 years, there are nuances that are different in small modular reactors.

Our hope is, as we go through this process which we budgeted right now in the neighborhood of $50 to $60 million, to go through design certification application, we'll also be able to enhance the advanced reactors going through, as you mentioned, the gold star or the gold plate of the NRC.

The CHAIRMAN. So you are paving the way for others, but I am sure that there are a lot of others sitting back and watching the NuScale process going forward saying maybe we do not want to be number two, maybe we want to wait until the process is a little more complete, having to produce these assessments or these analyses every step of the way.

Mr. Kuczynski, you have mentioned that Southern is looking into a more innovative licensing process. What are you recommending?

Mr. KUCZYNSKI. Yes, let me clarify and thank you for that question. I'll tag onto what John has to say.

In my opening remarks I talk about light-water technology, and the SMR that NuScale is building is a light-water technology, just a different type. So the existing regime is probably more suitable than it will be for advanced reactors.

And so, my comments are around what we are all calling now, modernizing the regulatory, kind of, regime. And through a lot of work with a lot of groups, we've all, kind of, landed on, I think, kind of, four cornerstones of what we think needs to be done in the regulatory arena.

One, it does need to be modernized to accommodate fuels that are completely different, designs that are completely different, outside of our knowledge base.

So there are really four things we're talking about. First being a more performance-based, meaning just set the expectation of what these products need to deliver and let the innovators figure out how to do it versus a really prescriptive approach. Second is bringing a more risk informed basis into the whole regulatory regime. The third one is really around a staged process and there's a number of reports out there that talk about trying to retire the technical risk in a more staged process instead of all the upfront investment without certainty of whether those kind of design concepts will be approved. And then the fourth one is really modernize a framework to be technology inclusive so that it doesn't focus just on light water technology.

What is very good though is as we progress in the SMR licensing activities, it really builds on our Generation 2, our advanced passive reactors, and that there are some generic issues that, we believe, will be resolved through the SMR licensing process that has a direct relationship to advanced reactors. Things like the emergency planning zones, containment, security, control room staffing, those are very good generic issues that, as John said, can pave the way for resolving some of these before advanced reactors need to be put together.

The CHAIRMAN. Alright, thank you.

Dr. Gilleland, I am well over my time, but if we have an opportunity to continue this, I would like to gain more information.

Senator Cantwell, please go ahead.

Senator CANTWELL. Thank you. Thank you, Madam Chair.

Dr. DeWitte, on behalf of Senator Alexander and myself, since you mentioned visiting museums as a young child, we invite you to visit the Manhattan Project, the B reactor in the State of Washington. I am sure Senator Alexander will extend his own invite to you.

The reason I am bringing that up is because, I think, you crystallized in your testimony the opportunity, how big of an opportunity, this was in the past. And, as you said, the giants that we stood on like Fermi.

My question is really to you if you want to weigh in, but also to Dr. Gilleland and Dr. Peters. This China market which I am a big fan of the U.S. trying a clean energy strategy just because they're 40 percent of the energy use. So anything we do together helps accelerate the deployment. They are now 73 percent dominated by coal. Could you talk about the market opportunity for the new reactors?

Dr. Peters, could you also talk about this new materials development, the research and development on new materials that would help us on our current, the development of new materials and how that helps us in reaching this market opportunity?

Dr. GILLELAND. Well, certainly working with China is a good thing for the United States because it is the huge market. It's the dominant market in the world. They're going to build tons of reactors and maybe someday hundreds.

Our motive, of course, was to work with China because it would make the most difference on the emissions front and they are set up to proceed with actually trying out these new reactors. So we have a joint venture which we're about to sign the CNNC to proceed with a prototype.

Our motive here also has been to keep the United States in the game, as it were, because I think we have this unique combination of capabilities between the national labs and our way of doing things rapidly and entrepreneurially and innovatively. And so, that combination is, I think, very, very powerful.

Now the agreement would involve us being equal partners. Whatever gets built in China can be built in the United States. In the 2030's when we must replace the coal plants, I would hope to see a wave of construction and activity back in the United States.

So it is a way for us to participate in the dominant market, make the biggest difference and equip ourselves to do the right thing in the United States.

Senator CANTWELL. The expertise of the labs helps us leverage this technology. So I want to bring in Dr. Peters here, that it is the labs that helps keep the supply chain in the United States, our skill level, our innovation, our technology. The supply chain would?

Dr. GILLELAND. Yes, there's nothing that keeps you more focused than trying to build something. Your dogma goes out the window because you're trying to solve real problems with a new type of reactor.

And we have engaged about 50 institutions in the United States in a very focused way. And most of that hundreds of millions of dollars has gone to those institutions. It is the building of the supply chain, a very advanced supply chain.

And I think in the future almost all of our advanced reactor activities will be international in nature. That's just the way the world is working right now. You look around the world where the energy is needed and where the emissions must be low and that's the message to us.

So we have to get in there and try to lead. That's what this is all about.

Senator CANTWELL. Dr. Peters?

Dr. PETERS. Yes, thanks, Senator.

So first and foremost let me completely support what you said that actually the national labs and universities in the U.S. are a differentiation and they remain a differentiation for us, as a country. Granted, manufacturing has wavered but the world still looks to us for that expertise. And so, that is an advantage that we still have.

A couple of examples, and this isn't just at INL, I would also mention Oak Ridge, Los Alamos and PNNL as being/having expertise. But two examples. In the fast reactor space, much like what TerraPower is developing, you have either a metal or an oxide fuel depending upon how you want to manage your safety case. But we still have world leading expertise in both of those fuel types. That's some of what is being brought to bear in the case of the collaborations with TerraPower as they're developing a metal fuel for a sodium fast reactor.

But also for high temperature gas reactors, Steve mentioned those TRISO fuels, so silicon carbide particle fuels. Oak Ridge has a world leading position there, and we support that as well.

So when you look at the advanced fuels for these reactors, the expertise sits at the U.S. national labs.

Senator CANTWELL. Thank you.

Thank you, Madam Chair.

The CHAIRMAN. Senator Cassidy?

Senator CASSIDY. Thank you, Madam Chair.

Mr. Hopkins, the SMRs, I was looking at your thing. You are still going to have it surrounded by 7.4 million gallons of water. I did not have time to do the math, but it still seems like although it is small and modular, it is only small relative to the current. But it is really still a substantial footprint, I presume?

Mr. HOPKINS. Actually sir, to your point, it's 7.2 million gallons. That's the ultimate heat sink that the module sits in. The actual footprint for a 600 megawatt plant is nominally about 32 acres. So you could expand that out to, let's say, 100 acres or 150 acres, but the footprint itself is only 32 acres.

So our sweet spot, we believe, in the United States quite frankly, is for coal replacement. So if you think of the majority of coal here in this U.S, it's been 300 to 600 megawatts. We're hoping, in fact our first project is actually for coal replacement, is to be able to put in with the existing infrastructure a 600 plant, 12-module facility.

Senator CASSIDY. Now let me ask, I live near Entergy River Bend Nuclear Power Plant in Louisiana. I went and toured it, and they literally have a paramilitary force on hand for security with video surveillance of the fences. Every now and then they catch a raccoon trying to sneak in, that sort of thing. Will that be required for the SMRs?

Mr. HOPKINS. We're currently in discussion with the NRC. If you look at the actual footprint of the plant itself, the reactor building is a rectangle box. And so we believe if you look at the minimal number of entrances and exits, we could reduce that security plan, but that's still something we're in discussion with the NRC. So our general belief is the answer is yes, we can reduce that security force. To what number? We don't know at this time.

Senator CASSIDY. But would it still require 24/7 security?

Mr. HOPKINS. Yes, sir.

Senator CASSIDY. With AK——

Mr. HOPKINS. Well, there's new technologies that are out that we're looking at right now that are non-lethal technologies that some of the labs are currently working on. So we're exploring different options. But we——

Senator CASSIDY. Mr. DeWitte, would you require that same sort of security footprint? Because I can imagine you are someplace desolate, if there is a desolate place in Alaska. I am sure there is not. [Laughter.]

Where you have this, sort of, small operation, but nonetheless it is radioactive material. Could somebody helicopter in, grab it and take off with it and make a dirty bomb?

Dr. DEWITTE. That's a great question.

First, that leads me to just think the work that NuScale is doing to help answer some of these questions and leading the way on these things because the discussions they're having will inform how we approach this. Because fundamentally, when you think about what we're doing and our size, it's very similar to research and test reactors that are at college campuses across the country. They're very safe, very secure. Their security intervention usually involves relying on campus or local police to get there in a certain time period because the amount of material is so small.

Fundamentally that's, kind of, the same approach that we believe we can take and still meet all the objectives, not to mention the fact that the material is very heavy and hard to get at and hard to deal with in order to actually divert to do something with it.

Senator CASSIDY. Okay.

Mr. Kuczynski, you had mentioned, you used the word "affordable." But what I understand about what Southern Company is doing with nuclear, it is quite unaffordable.

I asked an energy company why they were not doing what you all were, and they said listen, we can make ten natural gas plants for the cost of what they are doing, and they are cheaper to operate. Now I have learned to say what I have been told, not what I know. So I say that not to challenge, but just to say what someone else has said. Your thoughts about that? That natural gas, the economics of nuclear, at least the scale that you are doing right now does not work with natural gas?

Mr. KUCZYNSKI. Yeah, I would respectfully disagree about the economics and our working. Essentially our project, when we certified it about six years ago, we anticipated a 12 percent rate increase to our customer base. Currently today we expect that to be less than seven percent, maybe around six and a half percent.

So this is a $14 billion project that's actually going to come in on a lower rate impact where customers, when we're completed, then when we start it, we think that's a phenomenal achievement. And——

Senator CASSIDY. I do not know how all this works in your business model, so again, I am asking not to challenge, but to learn.

If somebody had built the equivalent number of natural gas pow-ered plants at today's fuel prices, would it have cost $14 billion to generate the same? I think you have 2,000 megawatts.

Mr. KUCZYNSKI. Yes, of course it would be less. But you make a big assumption about today's fuel prices. If you take a levelized ap-proach, equalize it over 40 to 60 years, there isn't anybody that would give you a contract on 40 years on gas prices.

So we diversify our fuel mix in an effort to do clean, safe, reliable and affordable energy for the long-term and you need a diversifica-tion of fuel supply. We cannot be all gas.

In fact, our fleet has now transformed itself from less than 20 percent gas to over 55 percent gas by just fuel switching. So our ability to switch fuels based on the economics is a tremendous ad-vantage for Southern Company, and that's why electricity rates are among the lowest in the country and reliability is the highest. And that's why we're able to attract economic development in our part of the country.

So we're huge proponents of nuclear. It's stable, strong, reliable. And this plant is going to be around for 60, likely 80, years, with very, very stable cost bases.

Senator CASSIDY. If I might ask though, the $14 billion that you are, for the 2,000 megawatts, how much would it have cost to do that with natural gas-fired plants?

Mr. KUCZYNSKI. Yeah, I'm not an expert at that at this point in time. I can get back to you afterwards.

Senator CASSIDY. Please.

Mr. KUCZYNSKI. But essentially you have to look at it not on to-day's fuel prices. And we've done the economic analysis with nine different scenarios and nuclear. Continuing our project is, by far, the most economic result to date for our project.

Senator CASSIDY. Thank you.

I yield back.

The CHAIRMAN. Thank you.

Senator Heinrich?

Senator HEINRICH. Madam Chairman, I want to thank you for this hearing. I think it is important that we look at advanced fis-sion technologies. I also want to suggest that it might be useful, at some point, to have a second hearing on the status of nuclear fusion research as well where there are some pretty exciting devel-opments, particularly around new materials developments and super conducting magnets at places like MIT.

I want to start with Dr. DeWitte. First off, where did you get your donuts? I think that is an important question.

Dr. DEWITTE. Johnson Donuts. [Laughter.]

Senator HEINRICH. Excellent, very good answer.

More to the point today, can you talk a little bit more about the design that you are pursuing, where you are in that process? And then what are the implications for things like fuel and spent fuel or waste, I should say, because you mentioned you are using spent fuel as your fuel source? So what does your waste stream look like and what are the challenges of dealing with spent fuel which is, obviously, highly radioactive?

Dr. DEWITTE. Sure. Thank you for the question, Senator.

So a couple things. We launched in 2013, and we've been working on this for a few years prior. And the reason we started on this very small concept is because we surveyed the field and saw that the economic opportunities of going small and starting off grid were very favorable, economically, for doing nuclear at these sizes, as well as the fact that it was a manageable approach to a new technology from a start ups perspective. Starting something small like that gave us a vector on how to tackle these issues.

So we've completed our confirmatory testing on what we're aiming to do with the system and completed verification and validation testing on full scale heat transport. We're moving into building an exact scale non-nuclear prototype later this year that will complete our transient testing on other things, much, similar, very similar, to the work that NuScale did several years ago building up the prototype testing plant.

One nice thing about us though is we're so small we can do everything at the exact scale. So that helps.

We anticipate submitting a license to the NRC sometime around 2018/2019, a license application, I should say. And what we would like to do is have our first reactor deployed in the very early 2020s. And we'd love to go as fast as we can to do that. But that's, kind of, the nominal targets we've set.

In terms of dealing with the fuel, we anticipate the very first reactors will be fueled with normal enriched uranium, low enriched uranium. Because of the difficulties in dealing with spent fuel, we don't really want to add that risk into the very first one. When I say risk, I mean technological risk, into the first reactor.

Senator HEINRICH. What kind of enrichment level?

Dr. DEWITTE. It's on the order of about 15 percent.

Senator HEINRICH. Okay.

Dr. DEWITTE. And mind you, these are very small reactors, so it's a small amount of fuel.

But we do, we are interested in opportunities to help with the plutonium disposition issue going on with, specifically related to the MOX plant in South Carolina.

Senator HEINRICH. Right.

Dr. DEWITTE. That would not be as much of a technological reach and it would prove out a lot that we'd like to do with spent fuel.

And there are some interesting technologies that we anticipate using to accelerate getting spent fuel into reactors for destruction and transmutation because what we do is we can fission all of the actinides over time. And what you're left with are fission products that normally have a half-life of about 30 years. So they're more or less gone in about 300 as opposed to the tens to hundreds of thousands with actinides.

Senator HEINRICH. Okay, thank you very much.

I want to get back to this issue that Senator Cassidy raised about general economics. Dr. Gilleland, I think you mentioned affordable nuclear power, and certainly Mr. Kuczynski, you walked through this a little bit. But I am having a hard time reconciling a number of information points.

I did a little bit of Google research this morning about this that is in the news, and one of those stories that came up was around two large reactors in Illinois that have light water reactors that have lost about $800 million over the past six years.

I am trying to figure out what kind of unsubsidized, levelized costs per kilowatt hour is everybody targeting to ensure that these advanced technologies are actually competitive in the marketplace?

For any of you really.

Dr. GILLELAND. Well certainly, as Bill Gates said all the way through from the beginning, if you can't afford it, it's all theory.

But the fact of the matter is in the case of the traveling wave reactor. Since you eventually do away with enrichment and you never need reprocessing, since the reactor would reduce the amount of waste produced by about a factor of five, since the basic fuel would be depleted uranium which is already mined, there's enough at Paducah to power the United States' fleet for hundreds of years. You add all those things up and you end up with a lower price of electricity. I'm not a great economist, but if you don't have to do something it's less expensive.

Senator HEINRICH. Sure, but——

Dr. GILLELAND. And so the——

Senator HEINRICH. There is going to be——

Dr. GILLELAND. Levelized carbon.

Senator HEINRICH. A target in terms of a cost per kilowatt hour that makes sense. I mean, we are seeing PPAs now at ridiculously low prices compared to what we saw a few years ago.

So I assume all of you have a goal for where you need to get to, to make sure that ten years from now as some of these technologies continue to come down the cost curve, that you are ahead of that.

Dr. GILLELAND. Yes.

And our levelized costs, I don't have those right in my head right at the moment. They're in the range of seven to eight cents per kilowatt hour, that kind of range, roughly speaking.

And we often compare to alternatives and since our system is so much simpler and since the waste produced is so much less and since the fuel is right now considered waste, that goes a long way. Some people confuse depleted uranium with spent fuel. Spent fuel is what's been used and it's now radioactive and is sitting around the country waiting for disposition. Depleted uranium is that vast quantity of uranium sitting behind enrichment plants. It's never seen a reactor. It's never been in one. Ninety something percent of all uranium mined is not useful as fuel.

And our objective was to take a look at Paducah and these other fuels of uranium and say, if you can burn that you're up a factor of ten in your fuel supply, it's already mined. There's no CO_2 in mining it. If you can burn it very efficiently with burn ups of 30 percent, several—an order of magnitude higher than present fission plants, then basically you're extending the fuel supply by about a factor of 40 to 50. And that's done without need for a proliferation prone processes such as reprocessing.

The CHAIRMAN. Thank you.

Senator Risch.

Senator RISCH. Well thank you very much, Madam Chair.

Dr. Gilleland, your understanding of economics is a lot like a lot of us Americans but you need to spend a little more time with the government. [Laughter.]

Senator RISCH. When the Government was shut down here they told us by not doing it, it was going to cost us more than doing it. So it gets very complicated when you get into government economics as opposed to just plain old common sense economics.

Dr. Peters, the advanced test reactor which is certainly one of our important facilities at the INL has been used for testing fuels for the Naval Nuclear Propulsion Program, and it has been very important in that regard. There has been some talk about how the ATR could be used to support the gateway for accelerated innovation for nuclear initiative. Could you talk about that for just a minute for us, please?

Dr. PETERS. Sure, thank you, Senator, thank you for the question. Good to see you.

So as you already said the advanced test reactor has been operating since the 60s, and its core mission to this day is to support the nuclear navy and that continues. And also note that we have a line of sight to this machine being operating as far as 2050, if not beyond, to continue to support that mission.

But there's already an existing part of the ATRs mission that's a part of the National Science User Facility that DOE funds that actually funds university researchers and industry researchers and other lab researchers to use some of the irradiation time at ATR to do work on our current nuclear energy system and also advanced nuclear reactor systems.

So really, yes, it will be a part of the test bed. It's a good example, actually. ATR and the High Flux Isotope Reactor at Oak Ridge are two good examples of test reactors that will be an important part of the GAIN test beds.

So and we're continuing to look to ways to, sort of, expand the capability of ATR to be able to be more responsive to industry needs.

Senator RISCH. Thank you, I appreciate that.

I recently met with the people who founded the Transatomic Power where they are dealing with these new molten salt reactors which, I guess, is kind of new to me and probably not to you people who work in this on a daily basis. But what can you tell us about that briefly, about this technology?

Dr. PETERS. It's an exciting technology.

Actually, the founders were at school with Jake, at the same time with Jake at MIT. So it's another example of some of these exciting startups that are emerging from the university community.

It's a molten salt technology. It operates at high temperatures. It's a very exciting technology. It's early in its development, but it's a perfect example of an early stage company that needs access through things like GAIN to capabilities at the laboratories to be able to do testing and evolve their design.

But, I mean, yes. There's—48 companies. Jake's being one, Transatomic being another that are out there with advanced concepts both fission and fusion, actually. And so the whole idea is to try to make sure that the labs are open to helping all of those companies at whatever stage of maturation they're in.

But there's reasons to be very excited about Transatomic. Southern's, for example, working with Oak Ridge about molten salt technology as well.

But when you look at the laboratories, the reactor technology sort of grew up through the laboratories in different ways. And actually the resident expertise in the laboratory system in molten salt tends to sit at Oak Ridge National Laboratory. They have a prominent capability there.

Senator RISCH. Thank you.

Last, every time I visit the INL there is always a discussion about new talent coming into the pipeline. There does not seem to be as much interest as there should be in young students wanting to take on nuclear physics.

Can you talk for a minute about the partnerships that the INL has formed with our universities in Idaho to try to nurture this and resolve this issue and move it along?

Dr. PETERS. Yes, sure.

So we have partnerships with, close partnerships, with Idaho State University, University of Idaho and Boise State, but in the nuclear energy area, particularly with Idaho State University, and the University of Idaho. And so we're working actively with them to bring students to the laboratory so that the students are not only getting training/education at the institution but also understanding doing research at the laboratory and a lot of back and forth between the university and the lab.

Also working with them to help devise their curricula in a more effective way to train the next generation. And that's a great partnership, but also, as part of the partnership that manages the laboratory, as you're aware, we have a national partnership. So we have other universities that are a part of that, part of the partnership that includes MIT, NC State, Ohio State, and University of New Mexico and Oregon State as well.

And so there's a national picture as well.

But I should say that, I mean, witness the guy at the end of the table down here. There's a change. The young people that are coming out today out of undergrad and grad school are looking to save the world and they're realizing that nuclear has got to be a part of it. I didn't feel like that was there, say 15 years ago, so I'm just really excited about it.

Senator RISCH. That is good to hear.

Thank you so much.

Thank you, Madam Chairman.

The CHAIRMAN. Senator Manchin?

Senator MANCHIN. Thank you, Madam Chair, and thank all of you.

I will start with you, Mr. Kuczynski, since you are in the utility business.

Do you agree with the EIA's forecast as far as an energy mix is going to be needed through 2040? And I guess, depending on whether the Clean Power Plan goes into effect or no, it does not. But they have, I think I can go a few figures. They had coal at 30, 32, 33 percent. You had natural gas at 30, 31, 32. Renewables at 18 and nuclear at 16. And even with the Clean Power Plan nuclear still stays at 16. That does not make sense to me. But do you agree with the forecast? Do you all see that?

Mr. KUCZYNSKI. We look at a number of forecasts whether they're out at APRI or EIA. We do know that we're headed to, I think, a carbon constrained future. That seems to be embedded in every, kind of——

Senator MANCHIN. Do you all have concerns as Southern about the reliability of this system?

Mr. KUCZYNSKI. We do not have concerns about the reliability of the system. We have an extremely robust system. Part of being in a regulative, regulated vertical we are able to invest significantly in our systems. So despite, you know, severe weather in our area we have tremendous functionality in our transmission distribution. We do believe nuclear is going to play a role, and we think some studies show nuclear playing a much stronger role. And that is why we're participating in this area of advanced reactors because not only do we have a deployable, large scale base load that we have solved the engineering, regulatory and almost all the construction risk already with the AP1000.

Senator MANCHIN. Let me ask——

Mr. KUCZYNSKI. It is ready to go. Advanced reactors could even expand us further. That's why——

Senator MANCHIN. Will advanced reactors, will you be able to ramp up and ramp down as power demands?

Mr. KUCZYNSKI. Many of the advanced reactor designs——

Senator MANCHIN. Because right now I do not think they do that, do they?

Mr. KUCZYNSKI. Are designed to do that. In fact, they could couple very nicely with the renewable energy sources.

Senator MANCHIN. That is what has been so attractive with the natural gas because gas is easy to ramp up and ramp down and the others are not quite.

Mr. KUCZYNSKI. Right. Each energy resource has its own features.

Senator MANCHIN. Okay. Does anybody else want to comment on that?

Dr. DEWITTE. I'll just add——

Senator MANCHIN. Go ahead.

Dr. DEWITTE. Advanced reactors changed the paradigm for load following and responding to the grid needs. I mean, our system operates fairly easily between 10 and 100 percent power and fairly quickly.

So it's an important feature, and it's going to change the economics, long-term, of what advanced nuclear reactors can bring to the table.

Senator MANCHIN. You all are very much involved as far as in the development of the new technologies as far as how nuclear will be used or could be used within the system, correct? I mean, all of you, I think, are in some form of that.

I am sure you have looked at other sectors and mine, in West Virginia, as you know, we do not have any nuclear power plants in West Virginia but we have a tremendous amount of coal-fired plants. We think that we do it as clean as possible and would like to even do it even more but we have no investment or no buy in at all from the Federal Government that is helping us to advance the technologies.

Have you been looking at some of that or do you see some advancement in that arena? Again, I know you all have done——

Mr. KUCZYNSKI. We, as a company, are investing in clean coal technology.

Senator MANCHIN. Coal.

Mr. KUCZYNSKI. With our Kemper facilities. You know, we are deeply embedded in trying to assure that that fuel source can meet our future energy sources.

Senator MANCHIN. Yes, and we are watching you all, but your cost overruns are pretty substantial.

Mr. KUCZYNSKI. Yes, it's true. You know, first of a kind big projects.

Senator MANCHIN. Right. That could be a cutting project for us to be able to use a dependable fuel, a reliable fuel such as coal for many years.

Mr. KUCZYNSKI. So as we get over those hurdles the technology will——

Senator MANCHIN. Southern, I believe, has coal in the mix for quite some time, right?

Mr. KUCZYNSKI. Oh yeah, we, that was our predominant energy source for many years. It no longer is. Gas is our predominant energy source.

Senator MANCHIN. Okay.

It is my understanding that China will add 23 nuclear reactors by 2020 increasing its capacity from 2 percent to 15 percent.

I am interested to know more about your company of TerraPower, Mr. Gilleland, and its agreement with China's National Nuclear Corporation to build traveling wave reactors, or TWRs? Can you tell me what's so attractive about this contract or how it is going?

Dr. GILLELAND. It's going well. We've been at it for ten years. We're a few hundred million dollars into the effort. We have used about 50 institutions in the United States, including national laboratories, in this development.

The State Department negotiated an agreement with us so we could freely exchange information with China, and that's going to be possible for other nations as well on the traveling wave reactor. The reason they did that for us is because eventually enrichment will not be needed and reprocessing will never be needed. That's where you take spent fuel and rework it again. People who analyze weapons proliferation risk say those are the two things that represent the greatest risk.

So our goal was to come up with a reactor which could be universally and ethically exportable, as one professor put it. We are about to sign an agreement with CNNC to have a joint venture——

Senator MANCHIN. Do you all plan on manufacturing in America or manufacturing overseas?

Dr. GILLELAND. It will be both places.

Generally, the first of a kinds are predominantly built here. Certainly the research and the leadership and the management of the joint venture will be the United States.

We plan to have the beginning of construction in 2018 or there-abouts with the first prototype plant going into operation——

Senator MANCHIN. Yes.

Dr. GILLELAND. In 2025 or 2026. The first commercial units will come a few years after that.

The important role that the national labs are playing here is in the materials, development and testing. So are the universities.

Senator MANCHIN. My time is running out, sir. I am so sorry to cut you short on that.

Dr. GILLELAND. Okay, sorry. Go ahead.

Senator MANCHIN. But if I could just ask one question?

Do you all believe that coal along with nuclear power is going to be needed for quite some time to guarantee the base load that is needed for this country? Do you all have any opinion on that at all? Anybody, just really quickly, if I may ask?

Mr. KUCZYNSKI. In our long-term study, coal still is a factor in our long-term generation mix.

Senator MANCHIN. As far as base load, you've got coal.

Mr. KUCZYNSKI. Correct.

Senator MANCHIN. Coal and nuke, right, for base load?

Mr. Peters, do you agree?

Dr. PETERS. Well yes, I think, but I'd also put in the plug for clean coal and continued——

Senator MANCHIN. Oh no, no, no, we're——

Dr. PETERS. Yeah, no, but innovation, including going to carbon capture and sequestration.

Senator MANCHIN. So basically for the reliability of the sys-tem——

Dr. PETERS. Right.

Senator MANCHIN. Coal is going to be needed the same as nuclear is going to be needed, correct?

Dr. PETERS. As a bridge to a future that, I think, looks quite different.

Senator MANCHIN. Okay.

Dr. PETERS. It's probably a very long bridge.

Senator MANCHIN. You are talking about beyond 2040, 2050.

Dr. PETERS. Yes, 2040, 2050.

Senator MANCHIN. Maybe you all could talk to the Administration and make them understand that. We would really appreciate it if you could.

Thank you.

The CHAIRMAN. Thank you, Senator Manchin.

Senator Alexander?

Senator ALEXANDER. Thanks, Madam Chairman, and thanks to the witnesses for coming.

Mr. Kuczynski, we have a couple of questions from Senators about cost. It costs TVA about $8 million to build 860 megawatts, and they built it in a year, I think. So it is costing you $14 billion to build 2,000 megawatts, right? That is a lot more expensive. The gas plant probably lasts 20 or 30 years. Is that about right? And the nuclear plant could last up to 80 years.

Mr. KUCZYNSKI. Up to 80, correct.

Senator ALEXANDER. And then you have the long-term cost of fuel which we do not know about, except we do know it has been a lot higher not long ago.

So my point is that even though at first $14 billion for 2,000 megawatts of nuclear does not compare very well with $2 billion for 2,000 megawatts of gas, if you take the length of time the plants might last and the importance of diversity in a big utility like yours, it does make sense to go with nuclear power.

Mr. KUCZYNSKI. We fully agree with, kind of, that summary. And 14 is full carrying cost, not overnight costs so——

Senator ALEXANDER. Let me ask you something else. Senator Heinrich mentioned the Illinois plants that were losing money. Exelon probably owns those plants. They are a merchant utility, right? And you are an investor-owned utility, is that correct?

Mr. KUCZYNSKI. Well they're investor-owned also, they're just in an unregulated market——

Senator ALEXANDER. They are in an unregulated market, and you are in a regulated market.

Mr. KUCZYNSKI. Correct.

Senator ALEXANDER. Exelon has testified that because of the size of the subsidy for wind power that basically, at some times, the wind producers in its region can give away its power to Exelon and still make a profit forcing Exelon to buy the wind power and not the nuclear power, making the nuclear power less viable.

Is the big wind subsidy that has been on the books for 23 or 24 years a deterrent to the expansion of carbon free nuclear power? Mr.

KUCZYNSKI. Yeah, I think the industry, you know, supports Exelon's position in those markets where it's not a true competitive market, and I think those of us that support nuclear believe we can compete in a levelized market that has an equal playing field.

I think in Illinois there is an over capacity. There's a lack of low gross and of subsidies and the massive growth in wind has just changed the dynamics of that market, and it's had unintended consequences with regards to those reactors.

Senator ALEXANDER. Let's talk for a moment about the amount of money.

Here we have a number of people who are engaged in clean energy research. We are talking about carbon free electricity that is reliable and at a reasonable cost. We have seen what has happened in other big countries, Germany and Japan, when they did not use nuclear power, and we saw the consequences on their manufacturing capacity.

But coming back here, we spend, according to the Congressional Budget Office, about $9 billion last year and this year on subsidies for wind, $9 billion. We spend about $5 billion on energy research, as a government.

Mr. Gates and others, including me, think we should double the amount of money we spend on energy research from $5 billion to $10 billion as rapidly as we can. That would permit people like you, or that would encourage people like you, to create new forms of clean, carbon free electricity. We might even find a way to have an economical method of capturing carbon from a coal plant or a gas plant which would be the Holy Grail, it seems to me, of carbon free electricity.

But even if you do not do that, my question to any of you is, wouldn't it be a better idea to phaseout this wind subsidy after 40 years, it is a mature technology according to the last Energy Secretary, and spend that $5 billion a year on energy research? We could instantly double the amount of money the United States spends on energy research if we did that. Wouldn't that be a better use of our money?

Dr. DeWitte. I'll jump in and say——

Dr. Peters. Senator, they're going to ask the national lab guy to answer that question. [Laughter.]

So I guess I'm going to beg off on the part about where the money comes from.

Senator Alexander. Well don't do that. I want an answer to the question.

Dr. Peters. Alright.

Senator Alexander. Are you going to continue to waste $5 billion a year on subsidies for that technology, or if somebody else wants to answer that, or are we going to spend it on encouraging people like you to create advanced reactors and small reactors or other forms of clean——

Dr. Peters. Senator, I think that if we increase the clean energy research funding, I have no doubt that we will unleash innovation and we will transform the energy sector.

Senator Alexander. Yes.

Dr. Peters. So increasing clean energy funding absolutely has to happen, but again I think it's outside of my purview to comment on where the money comes from.

Dr. Gilleland. Being a very unpolitical guy, the answer is yes.

Senator Alexander. Thank you.

Thank you, Madam Chairman. [Laughter.]

The Chairman. Dr. DeWitte, did you want to jump in there?

Dr. DeWitte. I was just going to say what John said, yes.

The Chairman. Okay, you got your answer.

Senator Franken?

Senator Franken. I would yield my time right now to Angus.

The Chairman. Thank you.

Senator King?

Senator King. First, Senator Alexander, my understanding is that the wind PTC that was extended in the last deal at the end of the year phases out over five years. I think what you are seeking is actually happening. That is my understanding. I may be incorrect about that.

Senator Alexander. Well that may be true. But in the next two years it is $5 billion a year, and that is exactly the amount of money we would like to have to double our energy research.

Senator King. I guess it is a question of what does phasing mean? It is phased out over a period of time. That was what was decided at the end of the year.

Anyway, I was very interested in Mr. Kuczynski, is it? You said you wanted to play on a level playing field. Are you advocating today that we repeal the Price Anderson Act and the nuclear industry should have to pay the full cost of insurance? That is what you said, I think, you want a level playing field.

Mr. KUCZYNSKI. Yes, I was more recognizing, kind of, the current subsidies that are in play and——

Senator KING. And Price Anderson is not a subsidy? Of course it is a subsidy.

Mr. KUCZYNSKI. We don't necessarily consider it a subsidy. It's not been utilized——

Senator KING. It walks like a duck, it is a duck, it is a subsidy. If you had to buy that insurance it would cost you a fortune, is that not correct? Yes or no?

Mr. KUCZYNSKI. The industry pulls——

Senator KING. No. Yes or no? If you had to buy insurance on the commercial market for your plants would it not cost you a lot of money?

Mr. KUCZYNSKI. I have not researched on exactly what that price would be.

Senator KING. Okay, so if you are not too worried about it then will you tell the Committee you think we should have repealed Price Anderson?

Mr. KUCZYNSKI. I think Price Anderson has been a valuable part of our energy strategy for 50 years.

Senator KING. I will say it is, but it is a subsidy.

Now in your $14 billion for your plant, were there any other subsidies?

Mr. KUCZYNSKI. We have utilized——

Senator KING. Do you receive subsidies from the Federal Government?

Mr. KUCZYNSKI. Yes, we have utilized, the way we believe subsidies ought to be utilized and for emerging technologies to get them restarted. So subsidies are——

Senator KING. How much of the $14 billion was Federal subsidy?

Mr. KUCZYNSKI. Well, we have not taken any direct subsidies. The only thing we have used to this point is loan guarantees which is technically not a subsidy, it's a financing mechanism that the Federal Government will be reimbursed for all financing costs.

Senator KING. Okay.

Mr. KUCZYNSKI. So we have not used any other subsidies, directly, for our projects.

Senator KING. I am inclined to agree with you on that. There were no direct subsidies of that $14 million. That is your testimony?

Mr. KUCZYNSKI. Correct.

Senator KING. Okay.

Dr. Cassidy is gone, but as I do the calculation, pretty straight forward, $7 million a megawatt for your plant. Two thousand into $14 billion.

Wind, which I know something about, $2 million a megawatt. Gas, between $600,000 and $700,000 a megawatt. So a huge differential.

I am not anti-nuclear. I like Maseratis. I just cannot afford them.

I do not understand any economic theory other than assumptions about natural gas prices that your power is going to be economic in the immediate future. Now if you do an 80-year calculation and you assume very high gas prices and no problems with disposal and waste disposal and all those kinds of things, I suppose you can make it work. But frankly, I just, again, I am not anti-nuclear, I just do not know how we can afford it.

Can anybody answer that question? The numbers do not work. Seven million dollars a megawatt? That is verses half a million or three quarters of a million, I mean, per megawatt for gas and $2 million for wind?

Dr. DeWitte. I think there's a couple things that are important to highlight there.

Senator King. I know the cost of gas is a factor, and that has to be. I clearly understand that.

Dr. DeWitte. Right, that and the fact you have capacity factors that do matter. Nuclear having excellent performance capabilities, delivering about 90 percent of their—capacity which is a big difference from what you see with typical renewables.

Senator King. Sure.

Dr. DeWitte. But it's also not nuclear verses renewables. I think that's an all too often, I think, pitted argument.

Senator King. No, no. That is why I put gas into the mix. I am just taking a range.

Dr. DeWitte. Right. And I'll say the other thing I would add to that is that advanced reactors do have the opportunity to fundamentally shift the economic paradigm of nuclear power. The advanced reactors usually don't operate at high pressures, use far less steel, far less concrete and they have a huge economic potential in terms of achieving costs that are more competitive.

Senator King. I agree with that. I think one of the great mistakes we made in this country in the nuclear industry was having each plant being an individual plant with its own unique design rather than a standardized design. Is this something we are moving toward is some kind of standardized design?

You are nodding. Is that? Could you?

Dr. Peters. The industry, the industry and NRC have already moved to that, Senator. And that would be part of the continued reform of the licensing process. When we go to advanced reactors the idea would be to not have it every design, you know, you have a design certification process up front, then you're simply licensing a site, constructing the same reactor.

Senator King. Right, and you can modularize.

Dr. Peters. Yeah, yeah.

Senator King. And standardize parts and those kinds of things.

You are now nodding. Is that correct? Is that where we're headed?

Mr. Hopkins. Yes, sir, that's the intent of the small modular reactor is people ask often why we limit our size to 50 megawatt electric because still a lot of our testing on seismic and etcetera, the height of the containment and the reactor itself, we wanted to ensure that it lent itself to a standardization of design so we could build these in a factory, both containment and the reactor. So what you have in the field is really a civils project, concrete to steel.

Senator King. Exactly.

Do we have any price? Do we have any per megawatt numbers on this new approach?

Mr. HOPKINS. Well, Senator Cassidy mentioned earlier about we do have a customer. They have to prove their economics before they go before their membership.

If you were looking today at Henry Hub prices of gas at less than $2 per million BTU, a levelized cost of engineering and combined cycle plant is nominally in the range of about $55 per megawatt hour.

Senator KING. Yes.

Mr. HOPKINS. We're right now, with small modular reactors at today's prices, based on the economics, about $72 per megawatt hour.

The question becomes if you look and it was mentioned earlier, in the West there's not a lot of gas. You have to bring the gas in. You have to go through the permitting process.

But the question becomes with LNG exports and other things and moderate production right now where it was over capacity. There was a lot of people moving in are now leaving the gas. What will the gas be in five to seven years?

Senator KING. Sure.

Listen, and once you do nuclear you have fixed your energy costs and your fuel costs. I understand that. That is true of hydro, wind, other. The difference is nuclear is base load. I understand that distinction.

The question is can we get that initial capital cost down to a place where it makes sense? I think that is the challenge and that is what we are talking about here.

Mr. HOPKINS. Well, what we're looking at right now is we have to be commercial viable for these to exist; otherwise there is no market.

And if you look at the small modular reactor, each of these units, of 12 of them, are independent. So you could put two or three and get them operational to stop—to start offsetting the cost of putting the additional plants in. So from a finance ability when you could put 600 megawatts for less than $3 billion U.S. Those are financeable.

Senator KING. Yes.

Mr. HOPKINS. And those are, you can put those on a balance sheet. And we currently have banks coming to us now, you know, saying here's how we think we can finance your project. So that's a big step.

Senator KING. It is a big step. Again, I hope that we can work, get to the place where we have economic capital costs and that then the technology can provide enormous fossil fuel free, carbon free energy.

Thank you.

Mr. HOPKINS. Absolutely.

The CHAIRMAN. We are going to move to Senator Franken here. Thank you.

Senator FRANKEN. Thank you.

Wait, what did you want to say, Dr. Peters?

Dr. PETERS. I was just going to re——

Thank you, Senator.

I was just going to re-emphasize that when you look at the research and innovation agenda for Generation IV reactors, a big part of what we're focused on is, in fact, addressing the cost, getting down the cost curve.

Senator FRANKEN. Okay.

Dr. PETERS. At the early stage as well.

So it is pushing the envelope on safety, burning the fuel more efficiently, but also going after design features that will help reduce capital cost because that is clearly a huge obstacle, I would say, to getting to commercialization.

Senator FRANKEN. Okay, that is enough. [Laughter.]

Senator FRANKEN. It has been five years since Fukushima, so it is important that we keep safety in the forefront when we are discussing nuclear power. I want to ask about these small modular reactors and advanced nuclear technologies and how they could potentially enhance safety if they are designed to operate without the need for external power to cool the reactors after all was loss of backup power from generators at the Fukushima plant that caused the cooling systems to fail.

Mr. Hopkins, can you give us an overview of the major safety concerns with traditional nuclear power and how SMRs improve safety such as by removing the need for backup power?

Mr. HOPKINS. Yes, sir.

Actually two weeks ago I was in Japan and had this conversation. If you recall what happened at Fukushima, it wasn't the earthquake, it was the tsunami that resulted in knocking out the electrical which therefore knocked out the cooling pumps and the plant couldn't cool itself down.

The small modular reactor passive safety systems. And this actually came about 15 years ago under a DOE program called Multiple Application of Small Modular Reactors. And the intent in the objective was to design a reactor with safety in mind. It wasn't about economics.

So even prior to Fukushima a lot of passive safety systems that were going on research at Oregon State University and Dr. Jose Reyes had to deal a lot with what the problems that occurred in Fukushima.

So our plan in passive safety, we refer to as the triple crown. If you were to have a station black out situation, the way that the reactor is designed it will cool itself down. You don't need operator intervention. You don't need additional electricity nor do you need additional water for this particular reactor to cool down.

And thinking this core is 120th the size of the large reactor. And so part of what we have at our Technical Advisory Boards that are made up of Senior Chief Nuclear Officers from 23 utilities and technical staff is to—and these are actual operators to look at those sequences. How would this reactor cool itself down? How can you circumvent a Fukushima event? And we believe the science is there and that's what we're currently working with the NRC with.

Senator FRANKEN. Okay.

Mr. Gilleland?

Dr. GILLELAND. Yes? It's the same.

Senator FRANKEN. What are the safety benefits of advanced nuclear designs like TerraPower's technology?

Dr. GILLELAND. It's a very similar answer.

The reactor is designed in such a way that the heat is efficiently conducted from the fuel out through the coolant, and it's because we use metal fuel and metal coolant.

Long story short, if you had a Fukushima there's no problem. You can have no internal or external power and the heat will be conducted to the outside world after the reactor, because of the way the physics works, shuts itself down. There's no need for a computer or a human to decide anything.

In addition, the mother of all accidents is that you also, not only lose that power for cooling, but you fail to put the control rods back in which is not what happened. This is worse than Fukushima. In that case the reactor also reduces automatically, its power to a very low level and can remain in that state indefinitely.

So that was the starting point for much of what our design effort was about. That first aspect of being able to use that type of reactor to shut itself down when there is loss of cooling was demonstrated at Idaho many, many years ago, the walk away reactor. So this gentleman's elderly people saw it done.

Senator FRANKEN. Okay, good.

Let me ask a question about the need for base load, because that has been brought up here. To what extent does advanced storage technology one, make the smaller reactors actually, maybe, a good idea since you do not need to necessarily be the base load? To what extent does better storage speak to that? And to what extent does better storage and real advances in storage decrease the need for the kind of base load that we have needed throughout our history? Dr.

GILLELAND. Well in my opinion, storage does not decrease the need for base load. It's a huge amount of energy, and load following which has been mentioned before would be very useful in these plants.

You wouldn't have to have storage. But at a certain point running a reactor steady state at full power for 90 percent of the time, that's a very economic way to operate a nuclear plant or power facilities.

But people usually use the term energy storage in the context of renewables which are inherently intermittent like wind and solar. There it would be helpful in the dispatching of that energy when you need it versus when the sun is shining or the wind is blowing.

Senator FRANKEN. That is what I am talking about.

Dr. GILLELAND. I beg your pardon?

Senator FRANKEN. That is what I am talking about.

Dr. GILLELAND. Oh, but you're still stuck with the fact that the solar constant is a constant and that even if you were to provide free energy storage you still have to think about the fact that you're going to have to rate the system to produce, in a short period of time, the energy you want to deploy over a longer period of time.

What that adds up to is a lot more acres, a lot more square miles going into solar panels or into sites for wind.

Senator FRANKEN. Okay.

Dr. GILLELAND. So there's a fundamental logic that says storage will help renewables but it's not going to get around the inherent problems associated with intermittency and the amount of energy density that's available from the sun or the wind.

Senator FRANKEN. Okay.

Does anybody else have an——

Dr. DeWITTE. I would just add that storage also can couple well with the nuclear power plants as well because you can charge those up at night and then discharge them in the day to either match up with renewables or match the curve. So there are opportunities for storage innovation improvements to partner well with nuclear technology.

Dr. PETERS. But the investment in storage is absolutely vital. You know, it doesn't replace.

Everything that they said was—I totally agree with, but innovation in storage is really an important thing to continue to support from the government.

Senator FRANKEN. I see a lot of nodding.

Dr. PETERS. Yes, really, really important to support.

Senator FRANKEN. Like for me too. I am nodding that my time is up. [Laughter.]

The CHAIRMAN. I agree on both counts. [Laughter.]

The CHAIRMAN. Storage and your time is up.

Let me turn to Senator Alexander, if you would like to pose a second question in the second round?

Senator ALEXANDER. Well thanks. This is very interesting and I thank you all.

On Senator Franken's point and Dr. DeWitte, I think that is very important. The disadvantage of nuclear power is that you cannot turn them on and off, but the use of them and the demand goes way up in the afternoon. So if we had a really good storage system, probably the greatest beneficiary of a really good storage system would be nuclear power because it produces so much electricity.

The second thing, on Dr. Gilleland's point, we use about a quarter of all electricity in the world in this country. So we're not going to run the country on windmills. I have said many times that is like going to war in sailboats when the nuclear navy is available. I mean, it is useful. It is helpful.

Third, on the subsidies, the Senator from Maine asked about the Pricewaterhouse. Well, the nuclear industry self-insures $3.75 billion which has never been used for accidents. So the first money comes from the nuclear industry. I think that would be important to point that out, Mr. Kuczynski, next time he gets asked about that.

On top of that, then the government might come in. But we come in from many emergency and disasters well before that. That would be our responsibility and it has never been used.

As far as the phase out of the wind subsidy, I mean, let's think about this. It has been going on for 24 years. The last energy secretary was a Nobel prize winner, and in testimony before this Committee he said it is a mature technology.

Now small reactors are not a mature technology. For the last five years we have been trying to pay for the government's part in helping that get off the ground. Advanced reactors are not a mature technology. That is where we might actually deal with climate change. We might actually deal with it there.

Here we are wasting $4 or $5 billion a year, and the point that it is phased out, that is a trick. That is a trick. It has been phased out more than a dozen times. That is called an extension. Now they have just extended it for a longer period of time and called it a phaseout, but it costs $4 or $5 billion a year. It is not $40 or $50 million, it is $4 or $5 billion. That is the amount of money that would double what we could spend on energy research.

So I think it is time for us to become rational about our energy policy, and rational to me means create an environment of government support through short-term support for new technologies. For example, there is a production tax credit for nuclear which, I guess, Mr. Kuczynski, you will take advantage of. But it is capped at 6,000 megawatts. The wind is uncapped. That is another big difference next time that question comes up.

So I favor, Madam Chairman, short-term support for new technologies and then they are on their own. The reason solar is about to be competitive is it does not have that kind of support. They have had some support, but nothing like these generous production taxes credits. As a result the cost of solar has been coming down, down, down, and it is about to get competitive as a supplement to the huge base load power that we need.

We need for the same thing to happen, not just with wind power, it needs to be on its own. I mean right now we have got a big company trying to build big towers to destroy the landscape in Tennessee where the wind blows 18 percent of the time. That is absolutely absurd.

TVA has said we don't need any more new electric base load power for 20 years, and we have taxpayers spending money they could be spending on clean energy research to build wind turbines in a place where they would just spoil the environment and where the power is not needed. That is really bad policy.

I would like to see, as far as NuScale's support, the support for advanced reactors. The whole idea there is that support will end and you will be competitive or you won't exist, right? I mean, you have said that to us in testimony if I am not mistaken.

Mr. HOPKINS. Correct.

Senator ALEXANDER. So, I think that is the approach, Madam Chairman, we should take with any of these new technologies where they are promising we should invest heavily in research, perhaps even support jump-starting a new technology like advanced reactors or maybe some, I mean, like small reactors, maybe some advanced reactor. But get out of the way, and then see what can survive.

Solar is about to survive. We hope NuScale will survive. We hope some of these new advanced reactors will have that as well. Let wind power survive too, then maybe some of these Illinois nuclear plants won't close because of negative pricing.

You have been very generous to let me extemporize here at the end, and I thank you for it.

The CHAIRMAN. Well, Senator Alexander, I know you have been occupied on the floor moving through education bills and our first appropriations bills, but we miss you in the Committee. Your contribution is not only important, but just a good reminder to us of the role that government should be playing as we help to facilitate this.

I think one of the things that I have enjoyed as I have learned more about these advanced reactor technologies is just the whole smorgasbord that is out there. We are not talking about one approach, one technology. There is a diversity now that I think we recognize. How we figure out how we encourage that rather than doing what we do around here, which is deciding who the winners and who the losers are going to be, and hoping that we bet on the right one.

So your comments about this, I think, are very important on the level of support that we should be providing at the Federal level. I want to ask one more quick question, and then we will wrap up. I appreciate everyone giving us so much time this morning.

This is directed to you, Dr. DeWitte, because I am very curious about the true potential for what you are describing with micro-reactors and the potential in remote areas. Whether it is a place like Alaska or you think about some of our islands and our territories. I have been going back and forth a lot with the folks in Guam about what we are going to do. We have military buildup there, but basically you are an island that is still powered by diesel. CNMI is still powered by diesel. Look at Puerto Rico, and the financial mess that they are in. So much of that comes to them because they have not been able to figure out how they deal with their energy. I look at some of these areas as just a perfect opportunity or en- vironment to have these smaller scaled technologies. But how you deploy them out—and let's use Alaska. Let's take an area like Bethel. You've got about 4,000 people out there. You are not at- tached by road. It is expensive in the first place.

How do you deploy? How do you deal with, you mentioned in re- sponse to, I believe it was Senator Cassidy, some of the issues about how you deal with the proliferation? You are sitting out there in Bethel, and I think you indicated that this is heavy, dif- ficult stuff to move. Realistically how could something like this work in remote, high cost areas with small populations?

Dr. DeWITTE. Thank you for the question, Senator.

That's exactly the market that we are targeting is to bring power to save money in places like Bethel where we would build probably between two and four reactors. And they work in a neat way be- cause they're designed in a containerized fashion such that we would ship basically two shipping containers nominally out per re- actor that would go. So in the case of four reactors, we'd have eight containers that would go up. In four of those containers are the re- actors themselves, the reactor module. We would then bury that in a hole that we'd dig, not very deep, about 20 feet deep. And then on top we would put the other container which contains the power—
—

The CHAIRMAN. We want to talk to you later about permafrost and how we deal with that, but yes. [Laughter.]

Dr. DeWITTE. Fair.

So in those situations we can actually mound up and it works above. Good point though. But and that actually adds a nice benefit for a couple reasons, permafrost itself, but that's a separate con- versation.

But anyway, you put the reactors there and then the process works that you can then tie up to either a microgrid solution or whatever the local grid system is there. And advances in power electronics actually enable us to do even more things in terms of grid matching and harmonizing with demand curves in small communities that are islanded from other grids effectively. And we would produce power for 12 years before refueling. And these systems use low enriched fuel, so they're not weaponizable material. They're also fairly small. I mean, you talk about fitting a shipping container. That also means there's not much fuel in there. But they're also not small enough that you can just throw it on the back of your pickup truck and drive away. They weigh 30 plus tons. And you don't have equipment left there that would be able to move them, as well as the emplacement that goes on top of them.

To the point that then the reactors themselves really aren't attractive, you know, targets to say to go after and try to get materials. Plus, from the intervention side and security, like Senator Cassidy mentioned and asked, you know, we have a staff. We'll plan on having security staff and personnel inside but it's also something we can respond to just like we have those plans in place for research and test reactors in other places. It's something that we can manage and definitely secure and make sure that it is not presenting a problem.

And the important thing though is that this provides a level of energy security and reliability that these communities have never really seen before, right? They no longer have to rely on diesel.

We've talked to some folks in certain communities that talk about bringing diesel in on dog sled. Those costs get super expensive very quickly. We eliminate those problems.

On top of that it's not just electricity we can provide, it's also process heat, right? So we can heat community centers or even local, if we tied into the infrastructure that is in place, district heating. And we could tie into that and supply for that.

And the important thing is that this saves a lot of money over diesel fuel, and it opens up, basically, larger portions of energy to be used by those populations to help overcome a lot of the challenges that they're stuck with which is, you know, constrained by limited access to power.

And in terms of operations, you know, we have these things operated, the small crew, because they're cooled purely by natural forces. We wanted to design something that's very robust such that it doesn't require much intervention or maintenance or monitoring. So you kind of operate this thing sort of like you think of an oil rig where you have crews going in and off, and that's basically how we provide that infrastructure and then produce power in these areas. And it gives us the opportunity to be able to put these nearly anywhere. So that's really the objective that we're trying to go after.

The CHAIRMAN. We have talked a fair amount about public-private partnerships and working with the national labs, working with universities. How much has been done with the Department of Defense?

They obviously have not necessarily a renewable mandate, but a goal toward reduced emissions on military installations. At one point in time the community of Galena, along the Yukon River, was actually looking at small nuclear as a potential for not only that community, but what, at that time, remained of a military installation.

How much discussion is going on with DOD for military installations, particularly in more remote areas whether it is Guam or whether it is Eielson Air Force Base?

Dr. DeWitte. We've engaged with different groups in the Department of Defense who've been interested in this. I think the big issue is they understand, generally, the need profile, but I think where we are and where we need to get to is to show that this is a mature technology that works. And I don't think the DOD is necessarily the right place to, well, let me rephrase that. The way they've looked at it is not necessarily the right place to start.

I think that can change as they see this going, and I think there's opportunities and partnerships between DOD and private industry like with what we're doing in possibly DOE in showing that this can work. But DOD has been hesitant to, I would say, take the lead on doing these reactor technologies because I think they still perceive them to be more on the developmental stage and not quite as ready for deployment. And we're hoping to change that perception very soon.

The CHAIRMAN. Mr. Hopkins, do you have any comments on that aspect of it or just the applicability of the small reactors?

Mr. HOPKINS. My understanding, Senator, is that in fact the DOD are looking now at small modular reactors. I think one thing that helped us all is the Administration's Executive Order on Alternative Energy last year came out and included SMRs as part of alternative energy which could open the brand with four Federal facilities.

If you look at your state in an area of Fairbanks, we have quite a few military installations, Fort Richardson, Fort Greeley, a large air force base. If you could work out a PPA with the local utility to provide the military installation reliable energy of which they need, not only for the military installations but also the surrounding community where the support staff is and where these people live. I envision that I could see applications for micro or small modular reactors in those types of instances.

The CHAIRMAN. I appreciate that.

Very interesting discussion this morning. I appreciate the contributions of all of you and your leadership in this area which, again, I believe very, very strongly, needs to be a robust part of our energy portfolio.

So thank you for your contributions.

And with that, we stand adjourned.

[Whereupon, at 11:59 a.m. the hearing was adjourned.]

APPENDIX MATERIAL SUBMITTED

———————

U.S. Senate Committee on Energy and Natural Resources
May 17, 2016 Hearing: The Status of Advanced Nuclear Technologies
Questions for the Record Submitted to Dr. Jacob DeWitte

Questions from Chairman Lisa Murkowski

Question 1: Can you please describe how advanced reactors, and specifically micro-reactor technologies, could work in microgrid applications, in remote communities, and for Department of Defense installations?

Micro-reactors enable a new paradigm for how energy can be generated. Advanced micro-reactors are well suited to generate power in micro-grids, isolated communities, and Department of Defense (DoD) installations. The electric power grids in these applications differ substantially from the large, national grid infrastructures in the continental US. Smaller grids don't have the same reserve capacity as larger grids, meaning a single generator can provide up to 100% of the power used in that infrastructure, which requires the generators to be flexible to respond to demand changes. Advanced reactors are capable of flexible operations, and can be well suited to meet fluctuating demand. This also enables them to be well-suited for coupling with renewable energy sources on the same infrastructure.

Micro-reactors have the potential to transform how these communities generate and use energy. These end users often rely on expensive diesel and fuel oil to generate their electricity because it provides a transportable, energy dense fuel that can produce reliable, albeit expensive electricity. However, this fuel must be transported to the site frequently to keep up with use, sometimes requiring icebreakers, planes, or even dog sleds to deliver the fuel. The costs and logistical challenges of doing this can be prohibitive, and prevent economic growth. Furthermore, some areas simply do not have access to energy at all. Advanced reactors can turn this paradigm on its head, because nuclear fuels are over 2 million times more energy dense, and significantly cheaper. This means one fuel load in an advanced reactor can provide decades of clean, reliable, and affordable power. This will save ratepayers money, enable new and expanded uses of energy, and drive economic growth. These benefits to off-grid communities also make micro-reactors attractive to DoD installation, remote mines, and behind-the-meter micro grids. These users need reliable, affordable power supplies that are immune to fuel supply disruptions. Micro-reactors are well-suited to meet these needs.

Micro-reactors can also bring clean, reliable, affordable energy to developing nations that do not have access to power at all, and do not have a large-scale power transmission infrastructure. Abundant energy is one of the most effective drivers of economic development and prosperity. Micro-reactors will play a crucial role in expanding global access to power, and the US has the potential to be world leader in developing and deploying this technology.

    a. What challenges do you face in deploying reactors of this scale?

        Some potential users in off-grid communities and facilities, such as those in Alaska are smaller companies or electric cooperatives that may have capital constraints to financing a micro-reactor. A micro-reactor might save

U.S. Senate Committee on Energy and Natural Resources
May 17, 2016 Hearing: The Status of Advanced Nuclear Technologies
Questions for the Record Submitted to Dr. Jacob DeWitte

ratepayers in these areas over 50% on their electric bill, but since nuclear is capital-intense upfront, this can be a challenge for some interested customers. Loan guarantee programs, and other vehicles that can defer those upfront costs will help these users adopt micro-reactors, and save their ratepayers considerably. Advanced manufacturing methods can also help reduce costs and shorten installation timeframes, which can enable faster and cheaper reactor siting.

There may also be regulatory challenges around certain operational aspects that Oklo is pursuing, but Oklo believes those are generally addressable within existing regulatory frameworks. Oklo does not see any significant impediments to licensing using existing regulatory frameworks, and believes the Nuclear Regulatory Commission (NRC) can license an advanced reactor in the near term. Oklo furthermore lauds the NRC for its work on developing non-Light Water Reactor (non-LWR) reactor design criteria. This will help advanced reactor developers commercialize more quickly.

b. How do you plan to overcome non-proliferation issues in remote applications of micro-reactors?

Oklo reactors are designed to be proliferation resistant. The reactors are sealed units that use low enriched uranium. They have long fuel cycles so they refuel infrequently, and will refuel at a central facility to minimize refueling in the field. This assures effective fuel handling and material accountability. Furthermore, Oklo reactors are up to 300 times more fuel efficient than current reactors so they use less fuel overall.

Advanced reactors are generally designed to be proliferation resistant, and can aid in anti-proliferation activities such as plutonium disposition. Fast reactors are particularly well suited for converting these materials that were once produced for weapons into peaceful, clean energy.

Question 2: Dr. DeWitte, what are the challenges in obtaining venture capital as a nuclear reactor designer/producer?

Raising capital for a nuclear reactor startup can have different challenges than raising capital for a more traditional Silicon Valley technology startup. However, the opportunity for advanced reactors is large, and there is an active and growing investor community. Investors appreciate the market opportunity and business model Oklo is pursuing. However, investors also consider investment risks around the challenges based on current electric power market pricing structures, challenges with siting first-of-a-kind units, and challenges with efficient and affordable access to DOE facilities that are crucial to reactor commercialization efforts. Oklo has not found perceived regulatory risks to be significant factor for investors. That said, startups must be wary of uninformed or misinformed investors, and the advanced reactor community should work to correct

U.S. Senate Committee on Energy and Natural Resources
May 17, 2016 Hearing: The Status of Advanced Nuclear Technologies
Questions for the Record Submitted to Dr. Jacob DeWitte

some perpetuating misunderstandings about certain aspects of the reactor commercialization, particularly the regulatory process.

A startup will be successful if it can make a product that people want, regardless of market, and Oklo is focused on making reactors people want. Demonstrating this, and meeting goals that de-risk technology development, de-risk the market, and de-risk deployment help a startup raise capital. Since nuclear startups in the current era are relatively new, identifying the relevant milestones can be challenging, but a focused team can execute and succeed in deploying reactors.

    a. Are there financial barriers to market that are specific to micro-reactors?

        Some potential users, like those in off-grid communities and facilities, may have limited access to capital. Since micro-reactors are capital intense, this can be a barrier that prevents these communities from purchasing micro-reactors. Programs that reduce these barriers by deferring the upfront costs to enable more communities to install micro-reactors will help a greater number of ratepayers save on their electric bills.

        The fee recovery basis for the NRC can hold the NRC back, and can prevent the licensing process from being as efficient as it could be. Changing the NRC fee model to a 50-50 structure would likely enable more streamlined reviews, more anticipatory policies, and an overall more efficient regulator.

Question 3. As a micro-reactor vendor, what do you believe the proper role is for DOE programs, like those sponsored through the Office of Nuclear Energy or through ARPA-E, in facilitating the development of micro-reactors?

All advanced reactor developers build upon the rich legacy of research and development of nuclear technologies done the DOE and the national laboratories over the last 70 years. The national laboratories are national assets, and are a strategic advantage for U.S. leadership in nuclear technology. The work they have done and are doing is critical to advanced nuclear commercialization efforts.

The recently formed Gateway for Accelerated Innovation in Nuclear (GAIN) initiative is a transformative step towards maximizing how DOE can support advanced reactor commercialization. GAIN provides a new and streamlined way for advanced reactor developers to work with the national laboratories to use their world class capabilities to support the commercialization of advanced reactors. GAIN's small business voucher program is an example of how the GAIN framework supports public-private partnerships, and this program should be expanded.

Recent advances in computational modeling and simulation enabled by DOE investments in research and development at the national laboratories and universities have particularly enabled advanced reactor developers to design and explore new technologies. In fact,

U.S. Senate Committee on Energy and Natural Resources
May 17, 2016 Hearing: The Status of Advanced Nuclear Technologies
Questions for the Record Submitted to Dr. Jacob DeWitte

Oklo engineers are able to use design and development techniques that were not possible ten years ago. Advances in modeling and simulation are one of the primary drivers behind the surge in advanced reactor commercialization efforts, and have made it cheaper than ever to design a nuclear reactor.

Additionally, many of these advanced reactor designs rely on data and insights gained from fuel and material development programs carried out by the national laboratories. Several advanced reactor developers are building upon the fuels and materials development work done at the Experimental Breeder Reactor II (EBR-II) and the Fast Flux Test Facility (FFTF). While the premature closure of these facilities in the 1990s was a national loss, these reactors demonstrated what advanced reactors can do. The data they generated is actively being used by several advanced reactor developers, and this data should continue to be properly curated.

Furthermore, the success of these test reactors show the need for a new fast test reactor in the U.S. Currently, if U.S. companies need to test materials or designs in a fast neutron environment, they must go overseas to countries with fast reactors, like Russia. Clearly this is not sustainable, and the U.S. needs a fast test reactor to support its growing advanced reactor industry so it can maintain and expand its leadership in advanced nuclear. DOE's investment in a fast test reactor would have a tremendous benefit on the nuclear innovation ecosystem.

DOE should continue to invest in universities, particularly in programs that grow the knowledge base in advanced reactors. Many of the entrepreneurs and engineers working on these advanced reactor technologies were supported by DOE fellowships or research grants, so clearly these programs are paying dividends. Maintaining this pipeline is one of the more important things DOE can do to grow U.S. leadership in nuclear technologies.

There are also opportunities to work on cross-cutting and enabling technologies that will support the commercialization and advancement of all advanced nuclear technologies, including micro-reactors. DOE should continue to support the national laboratories that are exploring and developing these technologies through investments by the Office of Nuclear Energy. Additionally, the DOE should support programs through ARPA-E that can maximally leverage ARPA-E's approach in areas like advanced sensors and control technologies. New sensors and automation technologies will enable new capabilities, and will help reduce both capital and operating costs.

DOE facilities can play an important role in supporting the commercialization of advanced reactors. National laboratories are well suited to host first-of-a-kind advanced reactors. DOE should streamline the relevant processes needed to realize this. DOE could also act as a customer, and direct its facilities to purchase the power from these facilities.

U.S. Senate Committee on Energy and Natural Resources
May 17, 2016 Hearing: The Status of Advanced Nuclear Technologies
Questions for the Record Submitted to Dr. Jacob DeWitte

While there are many things that DOE can do to assist with advanced reactor commercialization, DOE should avoid activities that send false signals to the market. For example, DOE should not engage in significant cost share grants with developers to design their reactors. These programs have historically had little, if any, success, because they have often been decoupled from market demands. These programs can also send false signals to the market if it is perceived that DOE is "picking a winner." Instead, DOE should invest in cross-cutting technologies that enable a variety of advanced reactor designs, spanning from mature through early stage technologies. This is also why it is key that any support from ARPA-E go to support of enabling technologies for reactors as opposed to commercializing a reactor type. DOE can also more effectively use taxpayer resources by investing in cost share programs with end users that are aimed at helping customers like utilities finance new reactors. This helps the market to select technologies, and helps draw customers to be early adopters. Ultimately, DOE plays a critical role in supporting advanced reactors and could even be the customer but must remain separate from the developer itself.

Question 4: What is the economic case for advanced reactors?

Advanced reactors have the potential to reduce the cost of electricity in the U.S. as well as in global markets. There are several factors why:

- These reactor designs often operate lower pressures than light water reactors, so they require less steel and concrete.

- Advanced reactors often operate at higher temperatures, which enables greater efficiencies.

- Advanced reactors can also benefit from factory based fabrication and manufacturing techniques, reducing on-site construction burdens.

- Many advanced reactor designs use fuel resources more efficiently and thoroughly, including used nuclear fuel from LWRs.

- The simplicity and operating nature of these technologies also offer to reduce operating costs.

- Furthermore, advanced reactor technologies enable reactor designs across a wide range of sizes, opening up new markets and opportunities, including new products such as process heat.

Altogether, these attributes can enable advanced reactors to achieve capital costs less than $2500/kW, and realize production costs less than $0.05/kWh.

   a. Specifically, what are your predictions for the levelized costs of your micro-reactor per kilowatt-hour?

U.S. Senate Committee on Energy and Natural Resources
May 17, 2016 Hearing: The Status of Advanced Nuclear Technologies
Questions for the Record Submitted to Dr. Jacob DeWitte

Oklo can save potential customers in remote areas such as those in Alaska over 30% and up to 80% on their electric bills in the near term. Further technology advances could enable Oklo designs to achieve costs less than $0.05 per kilowatt-hour.

b. In what markets do you believe your product will be economically competitive?

Micro-reactors can be deployed in a variety of settings, opening new opportunities for markets and customers. These include:

- Behind-the-meter microgrids
- Remote and off-grid communities and industrial facilities such as small cities and mining operations
- Mission-critical power users like data centers
- DoD facilities

Ultimately, Oklo is developing technologies with the aim of reducing the cost of energy worldwide, and strives to achieve cost parity with traditional on-grid power supplies.

c. Must markets be changed to allow for your reactor to be economically viable, in ways that attribute monetary value to the benefits of advanced nuclear power on the grid? If so, how?

Markets do not necessarily need to be changed to enable our design to be economically competitive, however policies that continue to disproportionally support other forms of energy can create inconsistent, unfair, and unsustainable market conditions that can put grid reliability and durability at risk. Current market policies do not recognize the reliability and clean nature of nuclear power, and do not recover the external costs of other energy sources. These conditions are putting existing nuclear power plants at risk of premature closure, which is undoing decades of progress made in expanding clean, affordable energy in the U.S. These policies are also incentivizing investment in technologies that reduce energy diversity in the U.S., making electricity costs susceptible to volatile fuel price changes.

Additionally, these market conditions do not reward the weather resilience that nuclear has demonstrated in hot, cold, cloudy, and sunny days, all while operating 24 hours per day. This is particularly important for states in colder climates where natural gas in increasingly used to for home heating such as New England. Limited pipeline capacity, and a heavy reliance on natural gas for power generation has made New England sensitive to particularly cold days, and supply disruptions. Nuclear power plants in the region have carried

U.S. Senate Committee on Energy and Natural Resources
May 17, 2016 Hearing: The Status of Advanced Nuclear Technologies
Questions for the Record Submitted to Dr. Jacob DeWitte

the load on those days, and have avoided the need for rationing. Market policies need to appropriately reward reliable power generators like nuclear.

The lack of these policies has led many utilities who were once leaders in nuclear technology to stay on the sidelines when it comes to new nuclear builds. The lack of fair and consistent market policies is deterring potential customers from participating in the advanced reactor technologies, which will help erode U.S. leadership in nuclear technology. Misaligned and out of touch market policies need to be addressed.

Reliable power is one of the most important factors to advancing economic development. The United States' tremendous economic growth, and phenomenal technical and societal achievements have been underpinned by access to reliable and affordable energy, largely driven by nuclear power. These trends will only continue, and the U.S. should strive to support its economy, and be a global leader in advanced nuclear technologies by supporting and enabling a flourishing nuclear industry.

Follow-Up Questions

Senate Energy Testimony of Dr. John Gilleland, TerraPower Chief Technical Officer

Question 1: What types of public-private partnerships have you utilized for the development of your reactor technologies?

TerraPower has numerous cooperative research undertakings with many public US Universities as well as Cooperative Research and Development Agreements (CRADAs) and Work for Others (WFOs) at National Laboratories including the Idaho National Laboratory (INL), Pacific Northwest National Laboratory (PNNL), Los Alamos National Laboratory (LANL), Argonne National Laboratory (ANL), Oak Ridge National Laboratory (ORNL), and the Y-12 National Security Complex. TerraPower is also part of an award from the DOE for the Advanced Reactor Concepts (ARC) R&D program for our MCFR technology.

Question 2: Can you describe the advanced molten salt reactor project that you are pursuing with a DOE cost-share along with other industry partners, including Southern Nuclear and Oak Ridge National Labs?

TerraPower's molten chloride fast reactor (MCFR) technology was awarded up to $40 million cost sharing from the U.S. Department of Energy (DOE) in a consortium led by Southern Company Services and including TerraPower, Oak Ridge National Laboratory, the Electric Power Research Institute and Vanderbilt University, for the purpose of research that is expected to further the development of the MCFR, an advanced concept for nuclear energy generation.

TerraPower believes that the MCFR technology has the potential to provide reactors with enhanced operational performance, safety, security and economic value. This Advanced Reactor Concept (ARC) project is anticipated to stretch over multiple years.

We are excited to participate in public-private partnerships related to advanced nuclear R&D. The scope of the multi-year award is summarized below.

- Integrated Effects Test (IET) to support the licensing, detail design, and operation of the Test Reactor
    - Demonstrate Key technologies including: Safety, Safeguards, and Security, Performance, Economics
- Key programmatic technologies in support of deployment.
- Licensing Strategies to support development program
- External reviews of technology readiness

Question 3: How does your pursuit of multiple advanced reactor technologies reflect your expectations of the market environment for these reactors in the coming decades?

TerraPower strongly believes that greater investment in energy R&D will help to address the need to deploy clean energy solutions to stem the tide of climate change. While we understand there are risks with new technology development, we take a disciplined, milestone-based approach to R&D and look for reference technologies that can benefit from modern day engineering tools. We are confident that further research and development will give rise to safer and more reliable nuclear power plants, where nuclear can play a larger role in energy production, beyond just electricity production.

Question 4: Can you please describe why TerraPower chose to build your first demonstration Traveling Wave Reactor in China and set up the agreement to test your technology there?

TerraPower considered a variety of options and markets, which included building in other countries that have active fast reactor programs. The U.S. and Chinese governments have executed an agreement to permit technical collaboration on the traveling wave reactor. China is an attractive market for several reasons: strong regulator, large market and interest in clean, non-emitting generation technologies, technically educated talent pool to support nuclear development, and a willingness to diversify and invest in advanced reactors.

Question 5: What is the economic case for advanced reactors? **[Previously Answered]**

    a. Specifically, what are your predictions for the levelized costs of your advanced reactors per kilowatt-hour?

The final version of the TWR is estimated to have a levelized cost of electricity between 5 and 6 cents per kilowatt-hour. This estimate is subject to a number of capital and operating assumptions of our advanced reactor as well as financing and investor return assumptions.

    b. In what markets do you believe your product will be economically competitive?

A 5 – 6 cents per kilowatt-hr price would be competitive in many electricity markets both here in the U.S. and internationally. The current struggles of nuclear energy in certain deregulated markets with historically cheap fossil fuels, no preference for CO_2 free generation, and highly subsidized/incentivized renewables are serious issues that require a comprehensive energy plan to solve.

    c. Must markets be changed to allow for your reactors to be economically viable, in ways that attribute monetary value to the benefits of advanced nuclear power on the grid? If so, how?

The Travelling Wave Reactor (TWR) or any future reactor must be designed to compete in a fair and equitable marketplace. However, in many cases the current market structure is not fair and equitable and a comprehensive approach is needed to ensure all forms of generation operate in a transparent and an open market place that reflects our shared goals and beliefs as a society. If we desire to limit CO_2 emissions from energy generation we should not select specific technologies but rather we should focus on the long-term goal of limiting CO_2 and then create market conditions that will incentivize this goal.

Specifically current energy markets fail to reward nuclear generation for several of its most prominent benefits to society. First, nuclear is a CO_2 free generation source (accounting for more than 60% of the U.S.'s emission free electricity) but unlike renewable energy there are no direct incentives available to promote new construction and the continued operation of existing facilities. Second, as renewables become a larger fraction of our energy generation portfolio grid stability will become increasingly valuable but our current system of capacity pricing structures fail to fully recognize these benefits and leave many nuclear facilities under rewarded for their role in grid stability.

Finally, nuclear brings the highest level of price stability and reliability of any electric generating source. Price stability and reliability are key both to the average consumer and to the U.S. business and industrial sector. The current shift of power generation to fuel sources with significant historical volatility (e.g., natural gas) creates a risk for power intensive industries throughout the country because even a short-term spike in electricity prices could put certain industrial facilities out of business and slow our economy. A nuclear power plant is a long-term hedge against electricity price spikes and both the current and future nuclear power plants in the United States deserve a market mechanism that recognizes the value of this hedge.

U.S. Senate Committee on Energy and Natural Resources
May 17, 2016 Hearing: The Status of Advanced Nuclear Technologies
Questions for the Record Submitted to Mr. John L. Hopkins

Questions from Chairman Lisa Murkowski

Question 1: What are your final – and greatest remaining – challenges to deploying your first commercial reactor?

Significant challenges remaining to deploying NuScale's first commercial reactor include:

- Expeditiously addressing requests from NRC during the DCA review cycles.

- Obtaining funding for final development and design.

- Completing and coordinating the Supply Chain.

- Assisting the customer through the Combined License Application (COLA) licensing process.

- Obtaining financing for the facility (e.g., federal support through a DOE loan guarantee would help.)

Question 2: One example of a public-private partnership is the plan you have developed with the Utah Association of Municipal Power Systems and the Department of Energy called the Western Initiative for Nuclear (WIN). You plan to build a NuScale SMR plant on DOE land at the Idaho National Lab, which will provide electricity to multiple western state utilities.

a. How did this partnership evolve?

Early work with the Western Governors Association (WGA) was directed at building a positive, receptive political environment for nuclear power, an effort buttressed by the supportive perspectives of Governors Otter, Herbert, Inslee, (former Governor) Kitzhaber, Martinez, and Mead. The effort revealed a political champion in the form of the governor of Idaho, Butch Otter, who, as the WGA chair, sponsored the initiative to create a policy on the future of Nuclear Energy in the West. This was followed by a WGA 10-Year Energy Vision sponsored by Governor Herbert. This plan included an explicit goal of accelerating the introduction of SMRs to the marketplace. NuScale announced Project WIN at the same WGA meeting at which the 10-Year Energy Vision was unveiled.

A parallel effort identified the Utah Association of Municipal Power Systems (UAMPS) as a public entity with a need to build carbon-free baseload generation assets, to replace retiring carbon-generating assets, in the West. A teaming agreement was signed among NuScale, UAMPS, representing municipal power systems in eight western states, and Energy Northwest, the owner and operator of the Columbia Generating Station nuclear plant in the state of Washington. The agreements were to develop NuScale's first SMR project with UAMPS as the plant owner and Energy Northwest as the plant operator. This was the first project announced under the WIN program.

In developing the first project (later named the Carbon Free Power Project or CFPP) NuScale and UAMPS worked to identify a candidate area with the correct resources and the local political and popular support for a nuclear mission. Southeast Idaho, specifically the community of Idaho Falls and the nearby Idaho National Laboratory (INL) proved to

U.S. Senate Committee on Energy and Natural Resources
May 17, 2016 Hearing: The Status of Advanced Nuclear Technologies
Questions for the Record Submitted to Mr. John L. Hopkins

be an ideal location. The long history of nuclear development at the INL ensured the availability of a large and superbly trained workforce for a nuclear project. The INL infrastructure and other resources quickly made the lab site itself a strong candidate for locating the new project and this prospect was viewed very positively by the INL management and by DOE. The local community has always strongly supported nuclear power and continues that support today under the leadership of Mayor Rebecca Casper. The local municipal utility Idaho Falls Power, led by General Manager Jackie Flowers, also strongly supports the project and is itself a member of UAMPS.

A 2013 award of cost-matching support for the NuScale Design Certification Application (DCA) and First of a Kind (FOAKE) engineering support from DOE was followed by a separate DOE cost-share award in August 2015 for the UAMPS COLA Preparation and Site Development. In 2016 DOE granted a Site Use Permit to UAMPS to support siting of an innovative SMR project within the boundary of the INL site. These actions have made DOE-HQ a partner in the private-public collaboration supporting this first commercial SMR project in the United States. This collaborative effort can now be said to comprise NuScale, UAMPS, ENW, Western Regional and State government supporters, strong and enthusiastic local political and business support, a DOE national laboratory, and DOE.

b. Can this be a model for further public-private partnerships?

Yes. Many circumstances had to align to create the environment for the UAMPS CFPP, but similar circumstances can be found or developed elsewhere. The efforts made on behalf of the CFPP can serve as a template for new nuclear projects elsewhere both domestically and also internationally. NuScale is using the template to further develop multiple projects within the umbrella of the WIN and also in several other areas of the United States.

Regardless of the rationale for the power generating need (such as, baseload electricity, desalination, process heat, load following for renewables support, or others), the confluence of state receptiveness, local support for the candidate site, a customer committed to the benefits of nuclear, and ongoing DOE support are very favorable ingredients for a successful project.

c. What are the greatest challenges remaining to the success of this partnership?

Several significant challenges remain to bring this collaboration to a successful conclusion, commercial operation of a 12 module, NuScale SMR power plant. The most important include:

- NuScale and its customer must maintain the schedules for the Design Certification (DC) and issuance of the Combined License. The need for power is predicated on having the SMR technology available within the window for replacement of coal-based-generating resources. If either the DC or COL suffers serious schedule delays, this could adversely affect both the real and perceived success of the partnership. This means initial submittals of the DCA and COLA must be complete and of very high quality and all answers to NRC review questions must be clear, responsive to the subject, and complete.

U.S. Senate Committee on Energy and Natural Resources
May 17, 2016 Hearing: The Status of Advanced Nuclear Technologies
Questions for the Record Submitted to Mr. John L. Hopkins

- Continued access to funding must be ensured to support the technology design and constructability efforts. Submission of the DCA does not bring an end to design. The overall plant design is about 40% complete at DCA submission. A great deal of work needs to be funded to finalize design and ensure constructability of the plant and efficient manufacturability of the modules.

d. What best practices have been learned so far in the development of this partnership?

Important lessons-learned from this first collaborative project include:

- Supplier and customer must communicate constantly and coordinate their interactions with the public, regulators, influencers, and decision makers. To effectively advance a specific project, the customer and the supplier must be in lockstep to explain the need for power and to explain the technology, respectively. Either party alone is not as effective

- Having a customer who will explain their analysis and act as an independent supporter of the technology is very helpful. This has provided an objective assessment of the case for SMRs in general and for NuScale in particular.

- Federal support for projects whether in the form of direct financial support, participation in R&D, or access to physical resources lowers the entry barriers for specific projects in terms of both cost and time.

- Involvement with a customer that has wind generation resources in its portfolio has provided real-world wind variability data in the region to be served by the proposed project. This data has expedited studies to demonstrate how the load following capabilities of the NuScale Power Modules can be effectively integrated with the actual intermittent production of a functioning wind farm.

- Ability to obtain favorable financing for municipal generating companies.

Question 3: What is the economic case for small modular reactors?

a. Specifically, what are your predictions for the levelized costs of your small modular reactors per kilowatt-hour?

NuScale's current estimate of the capital costs for construction of a 12 module NuScale SMR power plant is $2,895 million. This is an estimate for a first-of-a-kind plant in 2014 dollars. This equates to an overnight cost of $5,078 per kWe (net). The source of this data is a plant cost estimate performed in 2014, which looked at over 14,000 line items, consumed over 10,000 man-hours, and used prices based on budgetary quotes for 85% of the plant equipment. Fabrication costs for the Nuclear Power module were informed by discussions with a large U.S. nuclear fabricator. Module costs included estimating the cost to fabricate/purchase each individual component for over 75 unique components. Cladding, welding, bolting, and gasket material and labor were individually estimated. Vendor quotes were obtained for forgings, valves, instruments, pressurizer heaters.

Based on the discussions for this cost estimate and subsequent inquiries with respect to capabilities and interest of U.S. nuclear fabricators, NuScale has concluded that the existing manufacturing capability and supply chain can support at least the first 24

U.S. Senate Committee on Energy and Natural Resources
May 17, 2016 Hearing: The Status of Advanced Nuclear Technologies
Questions for the Record Submitted to Mr. John L. Hopkins

nuclear power modules without new manufacturing resources. Cost reductions from lessons learned are anticipated but reducing these lessons to mass manufacturing are not required for the first plant to be economically viable.

In addition to capital costs, calculations of the Levelized Cost of Electricity (LCOE) includes several other cost categories. These are owner's costs, fuel and fuel waste costs, operations & maintenance, and decommissioning. Owner's costs typically include HR and management infrastructure, central office, COLA, permits, NRC and ITAAC inspections, and legal fees, switchyard, owner's project development costs, and owner's post-COLA engineering services. A commercially financed plant will also have costs for taxes, including property taxes.

For municipally financed plants, the LCOE was calculated for a 40 year project life with 100% debt, financed at 4.12%. The LCOE for a municipally financed plant was calculated at $80/MWhr.

For commercially financed plants, LCOE was calculated for a 40 year project life with 55% debt, financed at 5.5%, and 45% equity at 10.0%. The LCOE for a commercially financed, FOAK plant was calculated at $106 /MWhr.

In both cases, the project life used in the calculations was 40 years, although the design life of the plant is 60 years. Also in both cases, the costs are for a first of a kind plant.

EIA has projected the LCOEs in $/MWhr for several competitive technologies for new plants first coming on-line in 2020. NuScale has attempted to reproduce EIA's assumptions for a NuScale plant. This analysis resulted in an LCOE value of $95/MWh. This creates a basis of comparison against the LCOE figures published by EIA. EIA published figures indicate that conventional coal units are expected to have an LCOE of $95; advanced coal units were projected at $116; and advanced coal with carbon capture and sequestration (CCS) were projected at $144. Conventional combined cycle natural gas units are expected to have an LCOE of $75; advanced combined cycle gas units were projected at $73; and advanced combined cycle gas units with CCS were projected at $100. None of the above estimates include an estimated cost for carbon emission other than installed CCS. LCOEs for non-dispatchable technologies were also calculated but must be interpreted carefully. Capacity factors for non-dispatchable technologies are simple averages for the respective technology, but the availability may not coincide with operator dispatched duty cycles. As a result, the LCOEs between dispatchable and non-dispatchable technologies are not directly comparable. The LCOEs projected for on-shore and off-shore wind were projected at $74 and $197, respectively. The LCOEs projected for solar photovoltaic and solar thermal were projected at $125 and $240, respectively.

b. In what markets do you believe your product will be economically competitive?

NuScale believes that power plants based on its Nuclear Power Module offer a viable alternative to natural gas for replacement of retiring coal capacity. Under the current market conditions, projects most likely to be receptive to new nuclear capacity will be in regulated states especially those for whom a diverse and stable portfolio of power generation is a priority. States where gas costs are high are also preferred.

U.S. Senate Committee on Energy and Natural Resources
May 17, 2016 Hearing: The Status of Advanced Nuclear Technologies
Questions for the Record Submitted to Mr. John L. Hopkins

Until electricity market conditions change, investor owned utilities in unregulated markets are not likely to view new nuclear as a competitive investment as can be inferred from the recent premature closures and planned closures of existing nuclear plants.

In unregulated generation markets that exist largely in the northeast, Texas, and California, due to market structures which reward generating technologies with low initial capital costs even with high marginal operating costs, generating technologies such as NuScale's will likely be unable to compete. Market structures in the unregulated markets do not adequately compensate capital intensive generation holders, such as nuclear units, for the upfront capital which needs to be recovered for their economic viability. This is despite the reality that nuclear generating units are competitive, if not the most competitive baseload technology, on a levelized cost basis over the long life of the assets. Unless the unregulated markets are reformed to allow for the adequate capital recovery of competitive forms of generation such as nuclear, it is unlikely new nuclear will be built in these areas and we will likely continue to see premature shutdowns of existing nuclear units suffering the same dilemma as new nuclear builds.

In those markets where new nuclear is economically viable, NuScale SMRs offer several advantages over traditional nuclear plants due to greater affordability and siting flexibility. Advantages of the NuScale SMR over traditional nuclear plants for replacement of coal plants include lower LCOE, lower construction and financing risk, improved safety, smaller emergency planning zones, and superior load following capability.

c. Must markets be changed to allow for your reactors to be economically viable, in ways that attribute monetary value to the benefits of advanced nuclear power on the grid? If so, how?

Conceptually, the environmental and air quality benefits of nuclear versus competing technologies are not properly valued. Benefits of competing technologies are exaggerated and promoted in manner that is not commensurate with long term environmental impacts.

Concerns with anthropogenic climate change and toxic air pollution issues drive the reduction in coal-fired generating capacity. However, when this capacity is replaced by natural gas units, the carbon dioxide generation per kWh is still about half that of the equivalent coal-fired capacity. Even this is underestimated, due to the failure to account for methane losses in drilling and leaks in pumping and transfer operations. Over the first hundred years after emission, methane has over 25 times the global warming effect of an equivalent weight of carbon dioxide.

Similarly, the environmental impact and footprints of wind and solar are underestimated in comparison to nuclear. While none of these technologies generate significant greenhouse gases, the land use footprint of nuclear is tiny compared to that of solar (120 times that of nuclear) and wind (700 times that of nuclear) and nuclear has none of the impact on raptors, bats, and critical habitats for desert species. Nevertheless, Renewable portfolio standards generally do not credit nuclear generation and government subsidies greatly favor wind and solar.

U.S. Senate Committee on Energy and Natural Resources
May 17, 2016 Hearing: The Status of Advanced Nuclear Technologies
Questions for the Record Submitted to Mr. John L. Hopkins

To help the market give proper value to the environmental benefits of nuclear power the following changes are proposed:

- If we are genuinely interested in combatting anthropogenic climate change, a carbon price, or some other disincentive that includes natural gas is essential.

- Renewable portfolio standards should be redefined as clean energy portfolio standards and include new nuclear generation. This definition is more consistent with the intent of the portfolio requirements, i.e., establish a minimum proportion of clean generating capacity, not to ensure market penetration for specific technologies.

- Subsidies for clean energy should be equal across the board for all clean energies on a per kW basis. Cost subsidies for solar and wind are not based on clear comparative indices. If one megawatt-hour of electricity is worth $10.00 to society because it came from a "clean" source, it should be worth that same $10.00 whether it comes from a wind, solar, nuclear, or hydroelectric facility.

- As previously discussed, regulated markets provide the ability for the large capital outlay associated with a new nuclear facility to be recovered over the facility life. However, unregulated generation markets do not provide adequate capital recovery signals. Therefore, unregulated market structures will need to be reformed for new builds of nuclear power units to be viable in the future.

- Assistance for the economic justification of the facility must be provided to key stakeholders due to the current historically low-cost natural gas environment (e.g., a PTC such as for wind and solar.)

Written Responses of Mr. Steve Kuczynski to Questions for the Record for the May 17, 2016 Hearing Before the United States Senate Committee on Energy and Natural Resources

Questions from Chairman Lisa Murkowski

Question 1: There are many types of advanced reactor technologies that have been supported by government-sponsored research and development, and many that are being supported by private investments. This includes differences in reactor size, core designs, coolants, and safety systems.

a. **What is the proper role for the DOE national labs in facilitating the development of diverse advanced nuclear reactors?**

Since the inception of atomic energy, the national labs have been at the forefront of innovation and technological advancement. This remains true today. I believe that the national labs should continue to play a central role in supporting research and development of advanced nuclear reactor technologies. This includes providing technical assistance for the safe construction and operation of advanced reactors, particularly in the test phase. Notably, within the existing regulatory structure, the national labs have the facilities and the licensing flexibility to accomplish more research and technical progress sooner than could likely be accomplished by private facilities acting alone. For instance, the Congressional Research Service recently authored a memorandum titled, "NRC Licensing of Proposed DOE Nuclear Facilities," which suggests that an NRC license would not be required to conduct research and test activities for advanced reactors at the national labs, even if those activities involve a partnership with a private entity. Similar kinds of activities at private facilities could take much longer to obtain the necessary NRC approvals. This is one reason that the national labs are very attractive for the current stage of advanced reactor research work.

b. **Why does diversity in advanced reactor technologies matter?**

Diversity in advanced reactor technologies will encourage much needed competition, resulting in safer, more efficient, and more economic reactor designs. At Southern Company, we believe the federal government should support advanced reactor programs without picking the ultimate winners or losers among competing technologies. Innovation requires competition. Within our own company, we take great pride in our culture of innovation and desire for step-up performance improvement in all facets of our business. We also believe that our federal government partners have the capability to create the right environment for innovation in the nuclear technology arena to flourish. This includes public-private partnerships that can harness the power of collaboration. We are pleased that Congress is devoting considerable time to expanding nuclear power in the United States, supporting research into advanced nuclear technologies, and reducing unwarranted economic and regulatory burdens on nuclear power generation.

c. **Over the next 20 years, how will these diverse technologies shape the advanced nuclear market in the U.S. and across the world?**

There is a vibrant market for nuclear power globally. Diverse technologies have the potential to speed this growth. If the U.S. can lead the way in nuclear technology innovation, the nation stands to benefit greatly from job growth in the nuclear supply chain.

Eventually, through research and development, testing, competition among various designs, a modern licensing framework, public-private partnerships, and other technological innovations, dominant technologies will emerge and attract investment to support design certification, and ultimately commercial deployment of advanced reactor technologies. The dominant advanced reactor technologies that emerge through innovation and competition will define the marketplace and supply chain for nuclear technologies for decades to come. These advanced reactor technologies will provide the nation with more options to meet U.S. and global clean air and energy diversity goals. American leadership is critical to ensuring that our nation is at the forefront of these opportunities. Furthermore, successful deployment of advanced nuclear reactor technologies will drive increased market demand for clean, baseload nuclear power.

Question 2: **The nuclear industry is investing significantly in advanced reactor development.**

a. **Can you describe some of the activities that are currently underway at Southern? Why do you believe strongly enough to be investing in advanced reactor technologies now?**

At Southern Nuclear, we have a talented group of engineers and nuclear energy experts who are currently focused on exploring advanced reactor technologies. We are also working in close collaboration with our industry partners through efforts like the Advanced Reactor Working Group.

Our most significant action to date has been participation in the DOE Advanced Reactor Concepts (ARC) program. On January 15, 2016, DOE selected a Southern Company-led proposal as one of two recipients of approximately $6 million for this year (up to $40 million over the next five years) to explore, develop, and demonstrate advanced nuclear reactor technologies. With non-federal cost-share contributions, this project represents up to $80 million in new advanced reactor research. Our partners in this public-private partnership are TerraPower, Oak Ridge National Laboratory, the Electric Power Research Institute, and Vanderbilt University. This project will bolster the development of molten chloride fast reactors (MCFR), an advanced concept for nuclear generation under development by TerraPower. In addition to the MCFR, we are also assisting in the development of modern Prismatic Block High Temperature Gas Cooled Reactor (HTGR) technologies, which are expected to be significantly more efficient than current operating reactors.

While Southern Company has not made any commitments toward construction of power plants with MCFR, HTGR, or other advanced reactor technologies, the potential for Gen-IV reactors is enormous. We know that significant new electric generating capacity will be required in the decades ahead to meet the nation's growing energy needs. Nuclear power is an attractive option for meeting this demand with reliable, affordable, clean sources of baseload electricity with zero emissions, and advanced reactors will be even more efficient, produce less byproduct material, have enhanced safety features, require an even smaller geographic footprint, come at a lower cost to customers, and be capable of using a broader range of fuel types.

In addition to the work we are doing with Gen-IV reactors, Southern Nuclear is also engaged with the industry's efforts to bring small modular reactors (SMRs) to market. In January of this year, Southern Nuclear and several other leading developers and potential customers announced a memorandum of understanding establishing a consortium called "SMR Start," which is designed to help accelerate SMR commercialization.

Southern Company is investing *now* in advanced reactor research because we believe innovation and leadership are key to ensuring affordable, reliable, safe, clean, and secure energy for our customers for decades to come. .

b. Additionally, as the chair of the Advanced Reactor Working Group, can you describe the industry perspective on the evolution of the advanced nuclear reactor market?

The Advanced Reactor Working Group (ARWG) has representatives from seven electric utilities and ten reactor design companies, and it is charged with developing an industry vision of a sustainable program to support the development and commercialization of advanced reactors, with the ambitious goal of achieving demonstration of multiple advanced reactors by 2025 and commercial deployment in the 2030-2035 timeframe. The ARWG is currently focused on developing and implementing an industry strategic plan for the development and commercialization of advanced reactors, developing proposed legislation to support that plan, working to implement a modernized NRC licensing framework to support advanced reactor licensing, and establishing a demonstration program for advanced reactor designs at a DOE site, utility site, or yet to be developed test center.

Within the ARWG, we believe that the growth in research and development in advanced nuclear reactor concepts is a promising indicator that there is a market for these technologies. A 2015 report published by Third Way identified nearly 50 companies investing in research and development of advanced nuclear technologies backed by over $1.3 billion in private capital.[1] In order to further encourage innovation in nuclear technology, however, it is imperative that the Federal government work alongside these researchers through public-private partnerships, provide for a regulatory framework that

---

[1] Samuel Brinton. "The Advanced Nuclear Industry," Third Way (June 15, 2015) available at http://www.thirdway.org/report/the-advanced-nuclear-industry.

readily applies to advanced reactor designs, and promote competition through a technology neutral approach.

Question 3: **The Nuclear Regulatory Commission license is the gold standard for reactors around the world. However, we have heard of the heavy burden associated with a first license for a new reactor technology. Advanced reactors have never been licensed by the NRC and there are challenges in building up proper advanced reactor expertise within the NRC to license these new designs.**

a. **Can reactors utilizing advanced designs be effectively licensed through the current Nuclear Regulatory Commission framework?**

While possible using an inefficient approach with many exemptions, licensing advanced nuclear reactors that do not use light-water reactor technology in the current regulatory framework is ineffective and has created a barrier to the engagement of the private sector, limiting the willingness to invest in advanced reactor technologies. In short, the current technical requirements, policy issues, and licensing processes are LWR based, assume a certain level of operating experience, and are very prescriptive. These aspects should be updated to fit advanced, non-light water reactors.

b. **If not, how should the licensing framework be structured to more effectively allow for the licensing of advanced nuclear reactors?**

First of all, safety must remain a key focus. The regulatory framework should be performance-based, risk-informed, employ a staged licensing approach, and allow for various kinds of technologies to be developed and licensed. The framework should also account for reactors using various fuel types. The staged, technology inclusive approach promoted by the Nuclear Energy Innovation and Modernization Act (NEIMA) introduced by Senator Jim Inhofe envisions changes to the NRC's licensing framework that will be needed for advanced reactors. Additionally, the upcoming report from the NRC's recently formed Advanced Reactor Regulatory Task Force promises to outline a licensing framework focusing on these same areas. Where current regulations are found to be appropriate for both the current LWRs and advanced reactor designs, those regulations should be adopted into the new licensing framework. Similarly, where small changes to the current regulations would allow for efficient licensing of advanced reactors, the changes should be made. Finally, certain aspects of the framework will need to be redesigned to fit the specific attributes associated with advanced reactor designs.

c. **How does U.S. leadership in advanced reactors help ensure that our country remains the gold standard of nuclear safety, and allow us to propagate safe nuclear technologies and standards around the world?**

The NRC sets the global gold standard for nuclear safety. As an industry, we exceed NRC standards through the hard work of many dedicated individuals in the operating fleet who strive to go well beyond compliance to achieve excellence in safety and operational reliability. U.S. operating data demonstrates this commitment to a culture of continuous safety and reliability improvements. By modernizing the regulatory framework for advanced reactors in the U.S., the NRC will be, once again, setting the benchmark for nuclear safety for the world – an example that will be followed by the international community. With other countries moving forward with non-LWR advanced reactor designs, the United States will continue to have a key role in leading the world in providing a real example of unparalleled safety culture.

Question 4: What is the economic case for advanced and small modular reactors?

a. Specifically, what are your predictions for the levelized costs of these types of advanced and small modular reactors per kilowatt-hour?

We have not yet identified with specificity these costs estimates. As a potential purchaser of these technologies, we are in the initial stages of reviewing a wide range of technical and economic feasibility factors. Nonetheless, because of the features and benefits of advanced designs, we would expect that the levelized costs would be even less for advanced and small modular reactor designs than for the existing generation of nuclear technologies.

b. In what markets do you believe these reactors will be economically competitive?

When commercial deployment is achieved, which could occur in the 2030-2035 timeframe, we believe advanced reactors will be able to compete economically with other energy sources. Today, new nuclear reactor projects compete well with other sources over the lifetime of the unit particularly in regulated markets. Unfortunately, under current market dynamics, nuclear reactors in certain deregulated markets are viewed as non-competitive with some sources such as wind energy, which continues to benefit from substantial subsidies that are unavailable to emission-free nuclear power. Advanced nuclear reactor designs have the potential to change the economic analysis as it relates to nuclear energy. Simpler designs will result in faster construction at lower costs. Modern designs will be more energy efficient and produce less waste, which will drive down storage costs dramatically. Many advanced designs will operate at very high temperatures allowing for the generation of process heat that can support other industrial operations. Additionally, small modular reactors provide the opportunity to place several initial units into commercial operation and offset the cost of later additional units on the same site. Finally, in what is likely to be a carbon constrained future market, nuclear power provides a dominant answer to providing clean, reliable electricity for our nation.

c. Must markets be changed to allow for these reactors to be economically viable, in ways that attribute monetary value to the benefits of nuclear power on the grid? If so, how?

We believe that the regulated markets prevalent in the Southeastern United States and elsewhere have provided for stable, low rates and reliable electricity, which has been essential to strong economic growth in these areas. Advanced reactors, which will be expected to operate for up to 60-80 years or longer, would be more competitive in regulated markets than deregulated markets, although deregulated markets could also benefit from deployment of advanced reactors.

Question 5: As Chair of the Advanced Reactor Working Group, what do you and the industry believe must occur in the near-term to place the nation on the right trajectory to achieve the demonstration of advanced reactors by 2025 and commercial deployment by 2030-2035?

It is important to note that this timeframe generally applies to large, approximately 1 GW reactors similar in size to the current operating fleet. Some innovative companies are targeting significantly earlier deployment due to smaller size and designs that leverage significant previous research, which may allow commercial deployment even sooner. Demonstration of large, advanced reactors by 2025 and commercial deployment in the 2030-2035 timeframe are ambitious yet achievable goals. This mission will require public-private collaboration, resulting in innovative policies, licensing frameworks, and regulatory structures that facilitate the efficient and predictable deployment of these new technologies and encourage private investment. Access to the expertise and infrastructure housed at the DOE national labs will be key to efficient deployment of advanced nuclear designs, especially in the early testing phase of the process. Additionally, in the near-term, a new licensing framework that supports the effective licensing of non-light-water reactor technologies is equally important. A modernized regulatory framework that effectively addresses the needs associated with licensing a non-light-water reactor will signal to the private sector that it can invest in research and development of advanced reactors knowing that the licensing environment does not favor a single technology, thereby allowing various kinds of technologies to be developed and licensed.

Question 6: Please further discuss the status of the Vogtle project:

a. Why was the Vogtle project important to Southern Company, in the context of fuel diversity, grid security, baseload power, and long-term customer rates?

The construction of Vogtle Units 3 and 4 represents a significant investment and the most important infrastructure project currently underway in Georgia. It is part of Southern Company's strategic plan for providing safe, clean, reliable, and affordable energy to meet the needs of our customers. In terms of fuel diversity, the Vogtle expansion is at the heart of Southern Company's commitment to maintaining a full portfolio of energy resources to best serve our customers and provide affordable electricity for the long-term. For example, Southern Company's diverse fuel portfolio has allowed us to go from producing about 20% of our generation from natural gas to about 55% to take advantage of lower natural gas prices. The Vogtle 3 and 4 project furthers our ability to be flexible to meet the needs of our customers based on rapidly changing market conditions. Additionally, by spreading our generation reliance across several generation options, we are able to deliver electricity to customers with an exceptional level of reliability.

Furthermore, high capacity factors associated with nuclear power generation (greater than 90%) make it perfectly suited to provide baseload power, which helps to ensure grid stability, voltage control, and other features essential to powering our economy. Vogtle Units 3 and 4 will provide stable, baseload power for up to 60-80 years that will promote fuel independence and grid stability.

As one of the project co-owners,[2] Georgia Power engages in detailed analyses in order to determine future generation needs in light of projected electricity demand. In concert with the Georgia Public Service Commission, Georgia Power evaluates its needs and the economics of meeting those needs through different generation resources in its Integrated Resource Plan (IRP). In 2004, Georgia Power submitted its IRP to the PSC, which showed a need to add baseload power generation to meet future demand. Georgia Power also notified the PSC that it would evaluate nuclear generation as an option to meet future resource needs. In its 2007 IRP, Georgia Power indicated that new nuclear generation performed well under a number of projection scenarios and presented the best cost option to meet its future baseload needs. At the direction of the PSC, Georgia Power issued a baseload request for proposals (RFP), which was conducted with the active participation of the PSC Staff and an Independent Evaluator. As part of this process, Georgia Power conducted extensive economic evaluations of nuclear power generation and other alternatives for providing baseload power. Various economic forecasts and models demonstrated the cost-effectiveness of constructing Vogtle Units 3 and 4 across a broad range of possible future costs and risks. Because it represented the best cost option for meeting future baseload needs of its customers, Georgia Power submitted its Vogtle Units 3 and 4 proposal to the PSC in the RFP. In 2009, the PSC issued a certificate to build Vogtle Units 3 and 4, which deemed the units to be the best economic choice for the citizens of Georgia while also acknowledging the risks associated constructing the first new nuclear units in the country in decades.

The Vogtle 3 and 4 project still presents the best cost option for our customers. It is an example of the various benefits of nuclear power, including zero emissions electricity production, high capacity factors and reliability, exceptional safety records, and long-term affordability without price swings common to other fuels.

b. What was the economic case for the Vogtle project and how has it changed since the project commenced, including the decreased rate increases you discussed during the hearing?

Our detailed analyses identified construction of new nuclear as the best cost option for customers in the state of Georgia. The original cost estimate of $6.1 billion included the total capital and construction costs certified by the Georgia Public Service Commission and the estimated financing costs at the time of certification. This initial estimate was made prior to the completion of all design work for the plant. Additionally, the regulatory process ensures a complete review of cost changes. To date, all requested cost changes

---

[2] In addition to Georgia Power, the other co-owners are Oglethorpe Power Corporation, the Municipal Electric Authority of Georgia, and the City of Dalton through Dalton Utilities.

submitted as part of the Vogtle Construction Monitoring process have been approved. Current forecasts estimate the total cost of the project to Southern Company to be approximately $7.88 billion. A significant portion of the increase is due to increases in the cost of money, while growth in the cost of construction is actually a fairly low percentage of the total, especially when compared to other large, first-of-a-kind construction projects. Despite this increase, the Vogtle project still represents the best cost option to Georgia customers. Original estimates regarding the Vogtle project anticipated a rate impact of approximately 12% to customers. Rate impact is now estimated to be less than 7% despite cost increases resulting from risks associated with the construction of large, first-of-a-kind projects. This estimate takes into consideration construction work in progress in rate base treatment, the effects of the lower costs of nuclear fuel compared to other forms of baseload generation, lower financing costs due to the utilization of the DOE loan guarantees, and the positive benefits of the production tax credits made available by Congress in the Energy Policy Act of 2005.

c. Does the Vogtle project include any direct or indirect subsidies, and if so, have these been necessary to the success of the project?

The Vogtle construction project has not received any direct subsidies from the federal government. However, the Energy Policy Act of 2005 provided loan guarantees for first-of-a-kind energy technologies, which we are utilizing at the Vogtle project. The loan guarantee is not a direct loan. It serves to reduce financing costs associated with the construction of a state-of-the-art nuclear power plant—the first of its kind and the first totally new nuclear power plant in 30 years.

We also expect to qualify for energy production tax credits made available by Congress in the Energy Policy Act of 2005 for innovative, first-of-a-kind energy projects.

Recently, former Senator Saxby Chambliss provided expert testimony to the Georgia Public Service Commission regarding why Congress chose to support first-of-a-kind projects like Plant Vogtle with loan guarantees and production tax credits. In his testimony, Senator Chambliss points out that Congress recognized the potential for additional costs and delays associated with licensing and construction of first-of-a-kind nuclear reactor technology and the lack of nuclear construction experience due to the hiatus in the building of new nuclear reactors. DOE loan guarantees and production tax credits were made available to new nuclear construction projects to mitigate these risks. Senator Chambliss concluded his testimony by stating that "addressing and overcoming these risks is necessary to realize the national goal of constructing the first new generation of nuclear unit plants we urgently need and which Congress has expressly determined to be a national priority."

Question 7: Are there any further challenges or opportunities regarding the advanced nuclear industry and Southern Nuclear's business model, discussed during the hearing, or otherwise, that you would like to further clarify?

Yes, thank you for this opportunity. First, I want to further address the question, asked by Senator Manchin, concerning our views regarding EIA's projections for nuclear power. We look at a range of energy projections and forecasts. The referenced projections do not appear to take into consideration possible advancements in Gen-IV reactor technologies, nor do those projections take into account potential fundamental changes in the regulatory arena. Future prices for natural gas and other energy sources are difficult to predict and subject to significant uncertainty and risks, which is one reason we believe fuel diversity remains an absolutely critical component of any sound energy policy, both for the nation as well as for our customers. I would also like to point out that some studies indicate nuclear power playing a stronger role in our nation's energy future than projected by EIA currently.

Second, I would like to clarify my remarks in response to Senator King's question regarding the Price-Anderson Act. Price-Anderson is not a subsidy to the nuclear industry. The federal government has never been called upon to pay a claim under Price-Anderson. Price-Anderson is provided at no cost to the government or taxpayers. Under Price-Anderson, owners of nuclear reactors pay premiums into a liability pool, accept liability for nuclear accidents, waive certain defenses, and share amongst the industry the costs of paying claims associated with nuclear accidents. The total liability pool, funded by the industry, currently exceeds $13 billion. This system has worked well for the nation since its creation in 1957 and was extended in the EPAct of 2005 through 2025. A reliable system of insurance and liability protection is crucial to the nuclear industry.

Finally, concerning Senator Manchin's question about the status of the Kemper project in Mississippi, I would note that Southern Company has stated that it is committed to developing real solutions to generate electricity using all of our domestic resources. This includes clean coal technologies. The Kemper project is a first-of-a-kind technology that will pave the way forward for America and the world for the use of an abundant natural resource in an environmentally friendly way. Costs should be reduced in subsequent plants due to the lessons learned from the initial construction project.

Questions from Senator Elizabeth Warren

Question 1: **In your testimony, you claimed that the nuclear industry supports Exelon's position that subsidies for wind power create an unfair playing field for nuclear. You serve on the Executive Committee of the Board of Directors of the Nuclear Energy Institute, and the Southern Nuclear Operating Company is a member of NEI. NEI has argued that, "The largest beneficiaries of federal energy incentives have been oil and gas, receiving more than half of all incentives provided since 1950." Do you agree with this conclusion?**

I am not personally aware of the data for this statement. I have asked NEI to respond to your staff.

Question 2: **Southern Company is currently constructing two new nuclear reactors at Plant Vogtle site in Waynesboro, Georgia.**

a. **The first of the reactors was initially scheduled to start producing power on April 1, 2016, more than one month ago. When do you now anticipate that these reactors will begin generating power?**

Commercial operation of Vogtle Units 3 and 4 is expected to begin in 2019 and 2020 respectively.

b. **Southern Company originally anticipated that it would spend $6.1 billion on its share of the project. How much do you now anticipate spending?**

The $6.1 billion figure includes the total capital and construction costs certified by the Georgia Public Service Commission and the estimated financing costs at the time of certification. This initial estimate was made prior to the completion of all design work for the plant. Additionally, the regulatory process ensures a complete review of cost changes. To date, all requested cost changes submitted as part of the Vogtle Construction Monitoring process have been approved. Current forecasts estimate the total cost of the project to Southern Company to be about $7.88 billion. A significant portion of the increase is due to increases in the cost of money, while growth in the cost of construction is actually a fairly low percentage of the total, especially when compared to other large, first-of-a-kind construction projects. Despite this increase, the Vogtle project still represents the best cost option to Georgia customers. Original estimates regarding the Vogtle project anticipated a rate impact of approximately 12% to customers. Rate impact to customers is now estimated to be less than 7% despite cost increases resulting from risks associated with the construction of large, first-of-a-kind projects. This estimate takes into consideration construction work in progress in rate base treatment, the effects of the lower costs of nuclear fuel compared to other forms of baseload generation, lower financing costs due to the utilization of the DOE loan guarantees, and the positive benefits of the production tax credits made available by Congress in the Energy Policy Act of 2005.

c. Why do you believe your initial estimates of cost and timeframe were incorrect?

The Vogtle 3 and 4 project is a first-of-a-kind nuclear construction project that Congress has identified as a national priority. In addition to facing challenges common to all large-scale construction projects, the Vogtle project has encountered other challenges associated with completing a first-of-a-kind technology deployment. The project has also faced challenges related to changing regulatory requirements. Although some of these first-of-a-kind difficulties have resulted in delays, the project continues to progress with approximately 60 percent complete. Further, despite cost increases due to these anticipated risks, the rate impact to customers for the Vogtle project is lower than initially anticipated. Vogtle Units 3 and 4 will provide clean, safe, reliable, and affordable energy for up to 60-80 years and represent the best cost option for customers in the State of Georgia even without considering the value that nuclear adds to the State of Georgia's future in light of pending environmental regulations and the project's economic benefits, such as job creation during construction and operation.

d. Please describe in detail all federal incentives and federal financial support for this project.

The Vogtle construction project has not received any direct subsidies from the federal government. However, the Energy Policy Act of 2005 provided loan guarantees for first-of-a-kind energy technologies, which we are utilizing at the Vogtle project. The loan guarantee is not a direct loan. It serves to reduce financing costs associated with the construction of a state-of-the-art nuclear power plant—the first of its kind and the first totally new nuclear power plant in 30 years. We also expect to qualify for energy production tax credits made available by Congress in the Energy Policy Act of 2005 for innovative, first-of-a-kind energy projects.

U.S. Senate Committee on Energy and Natural Resources
May 17, 2016 Hearing: The Status of Advanced Nuclear Technologies
Questions for the Record Submitted to Dr. Mark Peters

Questions from Chairman Lisa Murkowski

Question 1: There are many types of advanced reactor technologies that have been supported by government-sponsored research and development, and many that are being supported by private investments. This includes differences in reactor size, core designs, coolants, and safety systems. I believe in allowing the market to choose the ultimate winners.

a) Why does diversity in advanced reactor technologies matter?

The advanced reactor industry is growing rapidly in the United States and around the world, and innovation in nuclear is proceeding at a pace reminiscent of the early days of nuclear power. There is a need for diversity in electricity generation, and private and commercial entities are currently pursuing and driving diversity in advanced reactor technologies.

Micro reactors can bring clean, affordable, and reliable nuclear power to areas which cannot support larger plants. Small modular reactors (SMRs) provide the opportunity to add additional modules as electricity demand increases. These reactors also provide flexibility to vary electricity output over a short timeframe to allow load following of intermittent resources such as wind and solar. Traditional-sized advanced reactor technologies provide clean, zero-carbon baseload electricity to meet utility-scale demand while using technology innovations to enhance safety, security, reliability and efficiency.

Nuclear technologies are also enabling technologies for a diversity of applications, including:

- Electricity generation,
- Integration of renewable/intermittent energy sources onto grids of various complexity,
- Space exploration,
- Naval propulsion, and
- Process steam or heat for industrial applications, such as water desalination, chemical processing, enhanced oil recovery, and production of synthetic fuels.

Each application has different needs in terms of operating environment (e.g., temperature), refueling, output and maintenance, and fuel cycle management, to name a few. A diversity of nuclear technologies is necessary to enable a diversity of use.

Ideally, the market would ultimately be the final arbitrator for the value of specific reactor technologies and their respective abilities to meet diverse needs and applications. Providing the necessary technical foundation for licensing and commercialization decisions requires unique facilities and capabilities that do not exist in the private sector. For the U.S. to maintain its leadership position in developing and commercializing beneficial nuclear technologies, a new culture of innovation built on strategic public-private partnerships is necessary to ensure that market decisions regarding technology development and licensing challenges are based on sound scientific and technical evaluation processes.

b) Over the next 20 years, how will these diverse technologies shape the advanced nuclear market in the U.S. and across the world?

INL has a vision for the vital role nuclear energy must play as part of future global energy systems, both near-term and long-term. The U.S. is widely recognized as a world leader in the development of advanced nuclear reactors, as demonstrated by the numerous cooperative research agreements and partnerships in place between the U.S. Department of Energy and foreign partners. Leadership, however, is earned, not granted, and other nations are pursuing a leadership role in nuclear energy development and markets. Private investors are not constrained by these boundaries and will likely invest where they will have the necessary R&D support and a clear and predictable regulatory environment.

The U.S. has an opportunity to maintain and enhance its leadership role. SMRs and advanced reactors provide an opportunity to re-establish the domestic nuclear industry – a key to global leadership. The U.S. is poised to reinvest in and regain domestic manufacturing and supply chain capabilities lost over the past 30 years. SMRs and advanced nuclear reactors can be entirely sourced in the U.S., creating new advanced manufacturing facilities vital for economic growth. This reinvestment also positions the U.S. to compete with a nuclear energy export market currently dominated by Russia, South Korea, Japan, and China.

The outstanding safety record of the U.S. nuclear power industry is a direct result of the groundbreaking research, development and demonstration of the technologies and safety systems used in the current power reactor fleet developed in the last century. This leadership position enabled the U.S. to set the terms and act as a key resource for other nations as they developed reactors for power and research, and to ensure the security and eventual return of fuels and materials that could otherwise be diverted.

The investments made in research, development, deployment and manufacturing to accommodate advanced nuclear technologies will similarly benefit the current light water reactor fleet, as well as energy storage and fuel development across the range of nuclear and non-nuclear fuels. Potential other applications include synfuels and biofuels that hold the promise of extending the lifecycle of coal and other fossil fuels. Advanced reactors that can run on used fuel and better utilize actinides in fuel hold the potential of easing the path toward closing the fuel cycle and reducing proliferation risks.

We must maintain the new paradigm in nuclear innovation and nuclear energy. This paradigm, as epitomized by the GAIN initiative, involves new ways of working with a diverse nuclear community that includes utilities, startups, large nuclear suppliers, government entities, venture capitalists and other investors, non-profits, non-governmental organizations and policymakers. The national laboratories are becoming an entry point, or portal, for utilities, large nuclear suppliers, entrepreneurs, and small businesses to access our facilities and staff expertise.

The emphasis on strategic partnerships to research, develop, demonstrate and commercialize advanced nuclear technologies provides an expedited path forward. GAIN recognizes the key role of strategic partnerships and must continue to grow and evolve based on feedback from our partners. The Utah Associated Municipal Power Systems' Carbon Free Power Project, which aims to license and commercialize a NuScale-Power-designed small modular reactor located on the Idaho National Laboratory site, is primed to serve as a successful example of this approach.

Advanced nuclear R&D will also benefit a range of energy applications. The development of optimized fuel cladding for innovative fuel types will increase our understanding of materials science and materials behavior in extreme environments. Research into advanced cooling systems will help develop better battery and energy storage systems. As we develop enhanced control systems for nuclear reactors, both LWR and advanced designs, we enhance our understanding of cybersecurity and protection systems for a range of critical infrastructure applications.

The innovation culture that is being shaped now will drive developments in nuclear energy and in overall energy approaches over the next 20 years. We are poised to engage in a fundamental transformation in nuclear energy at the national and international levels. To realize this transformation, the federal government and the national laboratories must demonstrate their merits as partners in this endeavor. The feedback we have received through GAIN indicates that strategic partners are seeking predictability and reliability in the foundational technical support and access to expertise – in other words, the dialogue and technical support underway now have increased trust and confidence, but the GAIN initiative must continue to grow and evolve based to meet national needs.

Question 2: **The back end of the fuel cycle is often cited by those opposed to nuclear power as a reason to not support new nuclear builds. Can you detail how advanced reactor designs contend with the challenge of nuclear waste?**

Generally, advanced reactor designs use different kinds of fuel and coolant, resulting in enhanced safety, greater efficiency and less waste. But the end result depends largely upon the specific advanced reactor and is very design specific. For example, using fast neutrons, a fast reactor design can work with recycled actinides efficiently to make electricity, thereby reducing waste. Fast reactors are in fact ideally suited to burn long-lived actinides that make waste disposition challenging.

Advanced reactors are operational in countries around the globe, including France, Japan, Russia, and China. The U.S. has not built an advanced reactor, not due to concerns with the back end of the fuel cycle, but largely because of concerns and uncertainty about the licensing process and cost. The GAIN Initiative is intended to help industry overcome some of those obstacles.

The nuclear industry has come up with many technical solutions for dealing with nuclear waste. Many, however, have not been implemented because of political and economic obstacles: a long and often burdensome licensing process, duplicative safety requirements and construction costs.

On April 7, 1977 President Jimmy Carter issued an executive order that ended recycling of spent nuclear reactor fuel in the U.S., citing concerns about proliferation. Countries such as France, however, have safely and securely recycled for decades, and U.S. companies have limited incentives to invest in advanced reactors without recycling to increase efficiency and reduce final waste volume. Recycling used nuclear fuel would allow U.S. nuclear power plants to take full advantage of the unused energy in the used fuel while reducing the volume, heat, toxicity, and long-term challenges associated with the byproduct. Even if American plants recycled fuel and smaller amounts of waste were generated, there would remain a need for a U.S. repository to store the residual waste.

Question 3: The Idaho National Labs is a powerhouse of great energy research today, but was the original cradle of advanced reactor technologies. 52 early prototype reactors were built at INL since it was established in 1949, demonstrating the basic methods employed in today's advanced nuclear reactor designs.

a) **Can you talk about the historical context of advanced reactor technologies, and INL's role in the development and deployment of these technologies?**

Established as the National Reactor Testing Station, INL hosted 52 reactors of various types that served various functions in this new emerging science. They included, among others:

- Fuels and materials testing, like today's Advanced Test Reactor and the original Materials Test Reactor. These reactors allowed researchers to determine the best materials to use in the creation of fuel assemblies and reactor components.
- Safety testing reactors, including specialized reactor facilities like the Transient Reactor Test Facility, Power Burst Facility, and Loss Of Fluid Test reactor that allowed the testing of reactor systems during severe accident conditions.
- Demonstration reactors that proved the possibility of various unique types of reactor fuel, cooling and support systems, such as the Experimental Breeder Reactor-II, the Boiling Water Reactor (BORAX) series, and Naval propulsion reactors.

In some cases these reactors were developed to demonstrate the viability of their original design concept and then served additional missions with some modification and upgrade. In other cases the prototypes served as training facilities for future operation teams.

b) **What kind of unique expertise does INL have today in advanced reactor technologies?**

INL's work focuses on various aspects of advanced reactor technology, particularly with the High Temperature Gas Reactor and Sodium Fast Reactor concepts that offer greater safety and efficiency, reduced waste, and a modular, easily-scalable design. Specific areas of focus include:

- Fuel and materials development and qualification for multiple reactor types,
- Thermal-hydraulic testing capabilities,
- Design and analysis methods, verification, and validation, and
- Licensing support.

In addition, the nuclear-hybrid energy systems group is focused on developing methods and systems for seamlessly integrating baseload nuclear power with intermittent sources like wind and solar. INL's nuclear cyber security group is focused on developing methods to incorporate cyber security requirements at the design stage. INL is also exploring advanced recycling flowsheet and safeguards development for long-term waste management options with partially- or fully-closed fuel cycles.

Question 4: The GAIN initiative is operating out of INL and intended to help the nuclear community navigate federal bureaucracy and provide it with support to move advanced reactor designs forward.

a) **Please provide more detail about how GAIN will be structured and how GAIN will facilitate DOE and the National Labs working with industry?**

GAIN is the organizing principle for relevant, federally funded nuclear energy research, development and deployment programs. GAIN is envisioned to serve as a single point of access to facilitate industry access to the expertise, facilities, materials and data hosted at DOE's laboratories and partner facilities. Various funding assistance, consulting, and support methods are being developed with the goal of helping innovative nuclear technology developers complete the research, testing, and prototyping necessary to validate their technology and bring it to market.

b) **What are the major challenges to GAIN's success? Are there legislative actions that are necessary for GAIN to succeed?**

The major challenges for GAIN's success are similar to those of the industry at large: technology development is a difficult and long-term process and predictability of funding plays a pivotal role. A stable and predictable funding environment for RD&D activities and infrastructure, combined with the flexibility needed to adapt the funding to the prioritized issues as the innovation process evolves, is key to GAIN's success. Additional challenges include uncertainties in the energy market created by outside economic and political pressures.

Legislative actions such as the Nuclear Energy Innovation Capabilities Act help demonstrate the bipartisan support of efforts to develop advanced reactors. Support for creating a new fast reactor testing capability within the US and for maintaining and improving nuclear R&D infrastructure will have a significant benefit in developing technologies with the ability to use fuel more efficiently and vastly reduce the volume and hazards associated with managing waste.

Taking further action to streamline and increase the flexibility of the licensing process will be of great assistance in assuring that new zero-carbon, baseload sources like advanced nuclear are ready to replace our aging current fleet when they cease operations. The application of GAIN-led research, development and demonstration efforts will help ensure continued operation of the current light water reactor fleet, which will serve as a generation and technology bridge to tomorrow's advanced reactor fleet.

Daniel B. Poneman
CEO of Centrus Energy and Former Deputy Secretary of Energy

Written Testimony to the Senate Committee on Energy and Natural Resources
"Hearing to Examine the Status of Advanced Nuclear Technologies"

May 17, 2016

Chair Murkowski, Ranking Member Cantwell and members of the committee, thank you for convening this important hearing and providing the opportunity to submit testimony for the record.

On the subject of innovative nuclear technologies, I wish to make three points:

First, advances in reactor designs are only part of the equation; we also need to focus on restoring American leadership when it comes to the nuclear fuel cycle.

There is no question that new reactor technologies on the horizon – from Small Modular Reactors to Molten Salt Reactors to innovative nuclear batteries – hold enormous potential. The grants that the U.S. Department of Energy awards in these areas support exactly the kind of public-private partnerships that we should be encouraging to ensure America's future leadership in this global industry.

But reactors are only one part of the nuclear innovation equation. As a nation, we need to be far more aggressive in regaining the leadership we once held in fueling nuclear reactors in the United States and around the world. America, once the exclusive supplier of nuclear fuel for reactors outside the Soviet bloc, now depends on foreign countries to fuel our own commercial reactors. The United States has also gone from being a leader in uranium mining to producing less than four percent of the world's supply. Our nation would benefit greatly from policies that support and encourage the development of technological advances and cost reductions in uranium extraction, conversion, enrichment, transport, storage, and disposal.

Our leadership in the nuclear fuel cycle has been the cornerstone of America's nuclear security policy for decades – providing the deep subject matter expertise that underpinned U.S. diplomacy in critical negotiations around the world, including in the Nuclear Suppliers' Group, alongside a reliable civilian fuel supply in exchange for strong non-proliferation commitments. But now the international fuel market is dominated by state-owned corporations, where governments rely upon their domestic fuel cycle companies as crucial contributors to their national security and energy interests. The United States, by contrast, has lost the ability to supply our own reactors as well as our allies with an assured source of enrichment.

With nuclear power continuing to play a pivotal role in providing reliable, resilient, carbon-free power in the U.S. and increasingly around the world, we cannot afford to cede any more ground.

Second, innovation isn't just about building new reactors.

The United States is only now emerging from a three-decade hiatus in building new nuclear power plants. We had 112 reactors in 1990. Today, we are down to 99, and several more reliable, efficient reactors stand at risk of premature retirement because, as a Nation, we have not been able to find a way to compensate them fairly for providing zero-carbon electricity with a high degree of reliability.

Consider also the fact that, after a period during which not a single new U.S. nuclear power plant was built, the U.S. nuclear industry actually produced 40% *more* electricity in 2014 than it did in 1990. This remarkable feat is a testament to American innovation in nuclear energy, as America's nuclear scientists, engineers, and operators learned how to run these reactors more efficiently, with greater reliability, and with ever higher records of safety. This is also true in the nuclear fuel cycle, where the domestic mining community now employs world-leading technologies for *in situ* extraction, while the United States has also developed the world's most advanced enrichment technology.

The unmatched reliability of nuclear reactors makes them critical to the performance of the power grid itself. In January 2014, for example, when a "polar vortex" gripped New England in dangerous subzero temperatures for days, coal plants with frozen coal piles and natural gas plants with frozen pipes struggled to operate. Natural gas prices spiked through the roof. Salvation only came as a result of the outstanding operation of the region's nuclear plants, one of which has subsequently been retired.

With advanced computer modeling and other important research under way at our national laboratories, we can continue to upgrade and extend the useful life of our existing reactor fleet. While this may not capture the imagination in the same way as other advanced technologies the committee is discussing today, it is no less important because we need our existing plants to bridge the gap until these new technologies can be demonstrated, licensed, and constructed on a commercial scale.

Third, to compete in the global race for innovation, we need to have sustained, stable, long-term policies.

America's leading competitors in the nuclear marketplace are all state-owned corporations established by governments in Europe, Russia, and Asia. These governments have made a long-term commitment to advancing nuclear energy – not only in the nuclear fuel cycle and protected domestic markets, but also in fostering a growing fleet of reactors based on their indigenous designs, and a workforce of scientists, engineers, craft labor, and many others who are steeped in nuclear technology and the strong safety and security culture that is essential to the health of the industry. Once a customer commits to a foreign reactor technology, their business and thousands of jobs may be lost to American companies for the life of the plant.

While decisions about what types of power plants to build are made in the marketplace, that market is shaped and influenced by federal, state and local government policies designed to ensure reliability and affordability for consumers. Yet these markets are generally structured in such a way that does not account for the most valuable attributes of nuclear energy -- such as its unparalleled reliability, the fact that it is generates no carbon, or the fact that it provides important protections to consumers against extreme weather events or a sudden spike in fossil fuel prices.

Reasonable people can and do disagree about the best way to correct these market failures, but there is little disagreement that they pose a serious threat to the kind of investment and innovation our Nation needs. This is particularly true in the nuclear industry, which has much higher up-front capital costs but much lower fuel costs than fossil fuel plants. Financing these capital costs over a period of 30 or 40 years is especially difficult when the policy environment is uncertain and unstable. We need a sustained, long-term approach to drive private sector investment.

Thank you again for your commitment to addressing these issues. As today's witnesses attest, we all agree that nuclear power can – and must – play a pivotal role in our nation's energy future. That future will be built on innovation.

Testimony of Christina Back, Ph.D.
V.P., Nuclear Technologies and Materials, General Atomics
for the U.S. Senate Committee on Energy and Natural Resources Hearing to Examine the Status of
Advanced Nuclear Technologies.
May 17, 2016

Chairman Murkowski and Ranking Member Cantwell, thank you for the invitation to submit testimony for your hearing today. My name is Christina Back and I am the Vice President of Nuclear Technologies and Materials at General Atomics (GA). GA is a privately held company providing high-technology systems with over 60 years of experience in nuclear energy starting with the TRIGA research reactor. I will describe what "advanced reactors" are, and what we believe may be appropriate issues for you to consider when developing public policy for encouraging the development of new reactor concepts.

We believe that it is important for our country to increase its use of nuclear energy because it is critical to maintain a diversity of energy sources and nuclear provides emission-free, baseload electricity. If we could make nuclear energy cost-competitive it would provide thousands of years of safe, clean electricity for our country. In addition, remaining the technology leader in nuclear energy is critically important to minimize foreign dependence and strengthen national security.

Unfortunately, because nuclear energy using existing technology is currently too expensive to be competitive, the U.S. nuclear industry is in decline. To reverse this decline, the United States must do what we do best – call on the ingenuity of our scientists and engineers to create new ways to produce nuclear energy safely, cleanly, and at a considerably lower cost.

We are very pleased that there seems to be increased interest in this effort as shown by Members of this Committee, attention from the Administration, and efforts from industry.

It is extremely important that the U.S. resist the temptation to rely almost solely on improving existing technologies that may be at hand and, instead, take the time to develop new leapfrog technologies that may make nuclear energy truly desirable. Thus far, the term "advanced reactors" has been used rather loosely, and can mean different things to different people. Some people consider it to refer to any non-light water reactor, such as a gas-cooled, sodium-cooled, or molten salt-cooled reactor. Others use it to refer to a new light water reactor, such as a Small Modular Reactor (SMR).

To establish the context, let's remember that, fundamentally, nuclear energy involves splitting an atom and using the heat energy released to turn a generator to produce electricity. At the end of the day, electricity is a commodity, and many consumers do not care whether it is made from nuclear fuels or from burning coal or gas, or from renewables; what matters is its cost. To provide that commodity in today's world, an "advanced reactor" must improve over existing reactors in the following 4-core objectives. It must:
- produce significantly less costly, cost-competitive clean electricity,
- be safer,
- produce significantly less waste, and
- reduce proliferation risk.

❖ GENERAL ATOMICS

We believe every worthy advanced reactor concept must address these 4-core objectives jointly. It is not sufficient to excel at one without regard to the others.

Now, I would like to discuss General Atomics' reactor concept, the Energy Multiplier Module or EM^2, as a way to illustrate what "advanced" can really mean. EM^2 was designed, from the beginning, to meet the 4-core objectives I just mentioned.

In the design of EM^2, GA gave serious consideration to risk versus payoff, and we chose to employ innovative design and innovative engineered materials to reach our goals. What makes it compelling to rethink advanced reactors now, is that in the last 30 years scientists have made unprecedented advances in understanding materials. It is now possible to engineer and manipulate materials for specific applications. Use of customized materials and technologies is what we chose to do for EM^2. This is what sets GA apart.

Now I will go through each of the objectives. First is cost. The drive to minimize costs led to the design of a much smaller reactor that could produce much higher power output per reactor volume than today's reactors. It also led to a push to higher efficiency, i.e., 50% more electric power from the same amount of heat. We do this by producing the electricity from higher temperature heat. This requires new materials.

Second is safety. For a radical improvement in safety, EM^2 uses engineered ceramic materials that are capable of working in higher radiation and higher temperature environments. The fuel is contained in materials that can survive accident temperatures over 2 times higher and would not be subject to failure like those in Fukushima. While challenges remain, our results so far have been promising. If they hold up, we will revolutionize this industry.

Third is waste. Minimizing waste products is linked to better fuel utilization. For EM^2, this is accomplished by the innovation of long-burn core physics and by higher conversion efficiency. Consequently, EM^2 will use only one fifth of the fuel and produce one fifth of the waste for the same amount of electricity than a current light water reactor.

Finally, fourth is non-proliferation. The innovative design of EM^2 keeps the fuel in the reactor for 30 years, without the need to refuel or reposition fuel rods. Less handling of the fuel, and tight security allowed by offsite core fabrication, significantly reduce proliferation concerns and lower operating costs.

As a guiding principle, we believe that to bring advanced nuclear power into the market, the cost of nuclear must be significantly reduced below the existing levels projected for new light water reactors. This reactor, if it performs as designed, would produce power at perhaps 40% lower cost than today's existing nuclear reactors and, because it also is a modular reactor built in a factory and transported to the site, it would require a much lower capital investment per unit of below $1.5 billion. Because it would be built in a factory, it also would reduce proliferation concerns and reduce licensing costs. It would be shipped to the site and installed within 4 years, again keeping costs down.

EM^2 is only one way to get to the right answer. There may be many other interesting ideas, and many if not most will involve designing NEW materials for nuclear applications. We suggest many of these new advanced reactor concepts should be looked into, and several funded at affordable levels of $5 to $10 million a year for at least 4-5 years before a decision is made to go to the next level, or to drop them.

Whichever are chosen are likely to involve radically new technology requiring upfront investments involving risk. Some of these investments may not pay off, and even those that are successful could require at least 10 years to make any revenue. While GA has already invested $40 million in the EM^2 concept, these commercial realities make it very difficult for any company to justify long lead development expenditures. So, if having future sources of cost-competitive, nuclear power is in the

✦ GENERAL ATOMICS

interest of the United States, the Federal Government will have to increase its support of nuclear energy R&D. And it will have to target it toward the development of advanced reactors using leapfrog technologies, rather than concentrating nearly solely on developing minor improvements in existing nuclear technologies.

We very much appreciate your interest in this subject, and this opportunity to submit our testimony for the Record.

www.ingramcontent.com/pod-product-compliance
Lightning Source LLC
Chambersburg PA
CBHW081154180526
45170CB00006B/2072